SHINGLE
STYLES

Innovation and Tradition in
American Architecture
1874 to 1982

SHINGLE STYLES

Innovation and Tradition in
American Architecture
1874 to 1982

PHOTOGRAPHY BY BRET MORGAN

TEXT BY LELAND M. ROTH

PRODUCED BY NORFLEET PRESS

HARRY N. ABRAMS, INC., PUBLISHERS

Director & Producer: John G. Tucker

Editor: Bret Morgan   Designer: Charlotte Staub   Copy Editor: Janet Byrne

Library of Congress Cataloging-in-Publication Data

Roth, Leland M.
Shingle styles : innovation and tradition in American architecture
1874 to 1984 / photography by Bret Morgan ; text by Leland M. Roth.
p.   cm.
"A Norfleet Press book".
Includes bibliographical references and index.
ISBN 0–8109–4477–4
1. Architecture, Domestic—Shingle style.
2. Architecture, Modern—19th century—United States.
3. Architecture, Modern—20th century—United States.
I. Morgan, Bret. II. Title.
NA7207.R68   1999   728′.0973—dc 21   99–14102

Published in 1999 by Harry N. Abrams, Incorporated, New York

The text for this book was composed in Bembo, a typeface based on one used by the humanist
scholar Cardinal Pietro Bembo in 1495. The contemporary version was initiated by
Stanley Morison and released by the Monotype Corporation in 1929.
The display type was set in Bernhard Modern, designed by Lucian
Bernhard in 1937 for the American Type Founders.

Color separations by Bright Arts (H. K.) Ltd.
Printed and bound in China by Imago

Frontispiece: The entrance to Naumkeag (1884–87),
Stockbridge, Massachusetts, designed by McKim, Mead & White

Opposite: A detail of the Kingscote dining room (1880–81),
Newport, Rhode Island, by McKim, Mead & White

Harry N. Abrams, Inc.
100 Fifth Avenue
New York, N.Y. 10011
www.abramsbooks.com

CONTENTS

FIGURE I. *"Old House at Newport, R. I.," rear of the Bishop Berkeley house, "Whitehall,"*
Middleton, R. I., 1728. Plate 45 from New York Sketch Book of Architecture 1 *(December, 1874).*

INTRODUCTION

The Shingle Style

It began with a photograph, a view of a nearly forgotten eighteenth-century building—the first such photomechanical reproduction of a building in the United States. The photograph showed the long, sloping rear of Whitehall, the Bishop George Berkeley house in Middletown, Rhode Island, built 1728–29. From this angle the house seemed to be covered entirely in wooden shingles, over the undulating surface of the roof and down the rear wall (fig. 1). This photograph appeared in the January 1874 inaugural issue of *The New York Sketchbook of Architecture*, a large-format publication aimed at architects as well as the general public.[1] Its editorial purpose was to promote a new approach to design inspired by the American place and the American past.

The publication of this photograph announced an important turning point in the development of American architecture: a shift away from the preceding ungainly original invention in contemporary design (seen especially in the wild contrasts of High Victorian Gothic buildings). Instead the editors of the *Sketchbook* would propose an architecture that declared itself to be wholly American by referring to the architecture of the Founding Fathers. In the four years following the appearance of this photograph, there would emerge a new approach to domestic architecture that capitalized on the lightness and flexibility of wood. The new approach introduced a spaciousness and clarity of form that were indeed truly American. This architecture would reverberate across the nation in the 1880s and 1890s, often influencing the emerging architecture of regions far from the Eastern seaboard. It would persist through a long twilight until it was rediscovered in the 1950s, finally acquiring its definitive name and becoming the inspiration for much Postmodern design since the 1970s. That architecture is the Shingle Style.

The man responsible for the publication of the photograph of the Berkeley house is assumed to have been the young Charles Follen McKim, soon to be one of the founders of the prestigious New York architectural firm of McKim, Mead & White. In 1874, however, he was almost unknown. As the inaugural issue of the *The New York Sketchbook of Architecture* went to print, McKim was leaving Henry Hobson Richardson's office in New York City, where he had been an assistant, to establish his own practice. McKim's text accompanying the photograph noted that, aside from presenting renderings and photographs of important new buildings, the editors also hoped

to be able to do a little toward the much-needed task of preserving some record of the early architecture of our country, now fast disappearing. For this purpose they would most gratefully welcome any sketches, however slight, of the beautiful, quaint, and picturesque features which belong to so many buildings, now almost disregarded, of our Colonial and Revolutionary Period. All drawings or sketches for publication . . . must be addressed to Mr. H. H. Richardson, 57 Broadway, New York.[2]

Since Richardson was likely absorbed in his burgeoning architectural practice, it is assumed that McKim was the *de facto* editor of *The New York Sketchbook of Architecture*. Certainly in the next few years McKim published several views of Colonial buildings in the *Sketchbook* and contributed articles advocating the close study of Colonial architecture for both its historic and design merits. This study of eighteenth-century buildings began to shape McKim's own work, especially when he was engaged to make additions and modifications to eighteenth-century residences in Newport, Rhode Island. As this early work makes clear, the origins of the inventive Shingle Style and the more derivative Colonial Revival are complexly intertwined.[3] At first, however, it was the more "quaint and picturesque" aspects of seventeenth- and eighteenth-century architecture that influenced McKim (and Mead and White, who soon entered into partnership with him).

Significantly, it was not the decorous Georgian front of the Berkeley house that was published in the *Sketchbook*; rather, it was the rear view, then in nostalgically ramshackle shape (fig. 2). Looking at America's ancestral architecture from this vantage point hinted at far more freedom of invention

FIGURE 2. *"Whitehall," Bishop Berkeley house, front facade. (Photo: John T. Hopf, courtesy of Joanne Dunlap and the Colonial Dames of America in Rhode Island, Whitehall Museum House.)*

than would have been suggested by the symmetrical and formal front elevation of the Berkeley house.

The rear view of the Berkeley house was important, too, in that it was a high-quality photographic reproduction—a gelatine transfer—not a halftone made up of a dotted pattern. The nuances of texture and subtlety of form could be seen directly by the viewer and did not need to be translated by an engraver, as had been customary for architectural publications for the previous thirty years. Linear images were needed to publish perspectives of proposed, unbuilt structures, but now such images could be photomechanical reproductions of the architects' own ink line drawings, revealing subtleties of rendering technique that had been lost when drawings had to be redone as engravings for publication. High-quality photographic reproduction of architects' drawings and sketches was to be critical in presenting the many subtleties of surface texture of shingled buildings, and in the dissemination of this new architecture across America.

McKim's own line drawings in the *Sketchbook* reveal that he was not adept at rendering shingled surfaces, but his good friend (and fellow assistant to Richardson) Stanford White certainly was. McKim published many of White's drawings, some of them original sketches of Colonial buildings and other renderings of projects in Richardson's office. One of White's published drawings depicted the Hasbrouck house in Newburgh, New York, of 1750 (fig. 3). Showing again the rear of an eighteenth-century house, White's drawing demonstrates his facility in suggesting the irregular textures of aged shingled surfaces.

McKim, Mead & White were part of a larger group of architects, all of them looking at America's earliest architecture and freely extrapolating from

FIGURE 3. *Colonial Jonathan Hasbrouck house, Newburgh, New York, 1750. Drawing of the rear by Stanford White, dated 1875. Ink on paper, 9" x 12". (Photo: L. M. Roth.)*

it to create an original architecture linked to the nation's past. In Boston, these architects included William Ralph Emerson, Arthur Little, and Robert Peabody. In the 1870s Peabody contributed numerous articles extolling the charms of Colonial houses to another new publication, the *American Architect and Building News*. Other such architects included Charles Alonzo Rich and Bruce Price of New York, Wilson Eyre of Philadelphia, and John Calvin Stevens of Portland, Maine. In the 1880s, their numbers grew to include architects in the Midwest and even California, all of them developing this new and informal American architecture, usually wrapped in continuous envelopes of shingles.

The emergence of the Shingle Style accompanied profound changes in the economy and American society after the Civil War. These influences are multifold, and include certainly the sense of liberation that is fundamental to the American experience, the sense that established patterns and traditions are not immutable but, rather, that each generation can define itself and its own aspirations. With the growth of industry came the emergence of leisure time, at least for some elements of society. Leisure time, and a life that allowed for leisurely pursuits, was rare in the agricultural economy that dominated American life before 1860. The ordnance and materiel demands of the Civil War, however, caused profound changes in just five years. The period immediately following the Civil War was one of even greater growth of heavy industry and manufacturing, the expansion of old wealth and the creation of new affluence.

The growth of industry required capital that was raised by stock. Stock was sold by investment bankers not only to other banks but also to the upper middle class, whose increasing numbers of engineers, chemists, metallurgists, architects, lawyers, and other professionals were necessary for the growth of industry. The products of industry were sold using advertisements, and the new advertising industry was born. What were considered luxuries in 1865 became commonplace articles of everyday life by 1920.

For the wealthy, and for many who were merely prosperous, weekends and summers now became times to get away from one's ordinary routine, or to flee the stresses of industrialized life, to seek recreation amidst the pleasures of nature in the mountains, the forests, and at the seashore. The years marked by the rise of the Shingle Style witnessed the introduction to the United States of genteel outdoor sports such as croquet, lawn tennis, and badminton. Now, women could engage in these sports in mixed company with men, accounting in large measure for the burgeoning popularity of these sports.

The pursuit of recreational leisure in turn required a fresh architecture that celebrated the newly won economic freedom, an architecture unfettered by strictures of the past, a modern architecture that was nonetheless connected to American architectural traditions that represented comfort, reassurance, and stability. That was why the photograph of the back of Whitehall, the Bishop Berkeley house, proved so appealing to McKim and the audience of

The New York Sketchbook of Architecture. The house was old, but it was looked at from a new vantage point, and through the new medium of photography.

So the Shingle Style emerged, called into being by the leisured classes, who desired an architecture that spoke of easy and carefree pastimes, an architecture that was not pretentious or boastful, that connected with an ancestral past but was not held in check by it. For a few bright decades, from 1879 to 1916, the Shingle Style flourished. It was an architecture of fresh spirit and unbridled expansiveness that stood in sharp contrast to the buildings of the not-so-self-assured, particularly those *nouveaux riches* dazzled by Old World traditions and hierarchies. In Newport, then a sensitive barometer of architectural fashions, the Shingle Style interlude was very brief indeed.

Consider, for example, the difference between the wholly original and picturesque shingled house called The Breakers, built in Newport by the Boston architects Peabody & Stearns for Pierre Lorillard in 1877–78, and the house that replaced it. In 1885 Lorillard sold the wooden house to Cornelius Vanderbilt; seven years later it burned to the ground. When Vanderbilt rebuilt The Breakers, he turned not to any of the prestigious architects skilled in the inventive ways of the Shingle Style but instead to Richard Morris Hunt, who provided a grandiose marbled monument of columns, arches, and tile roofs, derived from Italian Renaissance models, especially the palazzi of Genoa.

Pierre Lorillard left the increasingly ostentatious life of Newport in the mid-1880s to develop a rustic resort community in the hills of New York west of the Hudson River, hard against the New Jersey state line. The new community was called Tuxedo Park, after a local Indian name. (When at the Park, Lorillard would dine in a novel informal evening jacket without tails, which, in recognition of its place of invention, came to be called a "tuxedo.")

FIGURE 4. *Bruce Price. Travis Van Buren house, Tuxedo Park, N.Y., c. 1884-85. From G. W. Sheldon,* Artistic Country Seats *(New York, 1885).*

Lorillard commissioned Bruce Price to design a series of commodious but modest shingled cottages as retreats for himself and like-minded wealthy "rusticators." No two of the cottages by Price were alike, but all shared a spirit of energetic invention. Price's cottage for Travis Van Buren (fig. 4) placed a large triangular gable with an elegantly stylized Palladian window over a cavernous arched entry, in which the shingles were pulled inward and upward and wrapped around the edges, as if distorted by some enormous gravitational force. In this tour de force of shingling, the horizontal courses of shingles wrapped around the entire cottage without any moldings to interrupt them. When published, Price's designs for Tuxedo Park were widely studied by other architects. They are particularly notable for their influence on a young Frank Lloyd Wright's design for his first personal residence.

"Old English" Influences

The sources of the American Shingle Style are many, but certainly one early influence was the work of English architects of the 1860s and 1870s. Beginning in the 1840s and 1850s, a number of English painters had begun to wander the Sussex countryside, sketching fragments of old Tudor houses because of their attractive picturesque qualities. Not long after, new cottages and houses began to be built incorporating characteristics of these venerable sixteenth-century houses. Out of this milieu came architect Richard Norman Shaw, who was to push this so-called Old English style the furthest.[4] The houses devised by Shaw combined expansive plans with rooms often grouped around a spacious "medieval" hall that served a variety of functions. Shaw's houses employed tall gabled roofs, with upper walls of half-timber and banked windows, and stone and/or brick on the ground floor. In a few instances (as in his Glen Andred), Shaw also used carved plaster panels,

FIGURE 5. *Henry Hobson Richardson. William Watts Sherman house, Newport, R. I., 1874-75. Perspective drawing, ink on paper, attributed to Stanford White. From* New York Sketch Book of Architecture 2 *(May 1875).*

FIGURE 6. *Henry Hobson Richardson. James Cheney house project, South Manchester, Conn., 1878. Aerial perspective, signed (lower right corner) "Stanford White del[ineator]/'78." From* American Architect and Building News *(May 25, 1878).*

inspired by the carved plaster work of Sparrow's house, Ipswich, built circa 1670. Shaw revived the use of tiles as wall sheathing, along with matching red roof tiles, achieving a deep richness and continuity of color. His large country house, Leyes Wood, near Groombridge in Sussex, 1868–70, is perhaps his best known work of this type and has often been reproduced, but equally important to American observers were other Shaw houses, such as Glen Andred and Hopedene, in Surrey, 1873–74. These latter two are significant for their emphasis on horizontal lines, a feature that would influence the work of many American architects. Shaw's use of stacked polygonal end bays, capped with a projecting polygonal roof, is an antecedent of Americans' much-used corner tower.

Most important for American architects, Shaw's houses were profiled in the British *Building News* (and soon after in *American Architect and Building News*), which made use of Shaw's own crisply beautiful ink drawings. Henry Hobson Richardson had access to these English and American journals (and thus so did McKim and White). Moreover, Richardson had in his office library the books of Shaw's published travel sketches. Richardson's principal delineator, Stanford White, studied Shaw's technique closely, making his own drafting style a continuation of Shaw's. This can be seen in a number of drawings of Richardson projects published in the *New York Sketchbook*, such as the perspective of the William Watts Sherman house in Newport, 1874–76 (fig. 5), and in the early issues of *American Architect and Building News*, which ran an aerial view of the James Cheney house, 1878, in whose design White may in fact have had a hand (fig. 6).

Even while architects like Richardson and the youthful McKim and White studied Shaw's technique, they could not help but note that Shaw was creating a new and modern English architecture by examining much older vernacular traditions. This point was doubly emphasized in 1876, when the English architect Thomas Harris based the three buildings of the official British compound at the Centennial World's Fair, held in Philadelphia, on similar Tudor and Jacobean sources. Harris's designs were published as well as seen by thousands of fair visitors.

American Colonial Influences

Influenced by Shaw's example, it was natural for McKim to look to American Colonial architecture as a reference point in developing a new American architecture. Having become acquainted with several Newport families, McKim spent considerable time visiting Rhode Island in the early 1870s, familiarizing himself with the rich architectural heritage that remained there. He hired a photographer to make records of some of these seventeenth and eighteenth-century Newport buildings, and the print of the Berkeley house shows how he used these views.[5] Then in 1877 McKim persuaded his friend Stanford White and two other close acquaintances, William R. Mead and William Bigelow, to accompany him on a walking trip of the New England coast—what William R. Mead would later describe as "our 'celebrated' trip" —specifically to study and record Colonial buildings. Although Mead (who would soon become a partner in the new firm) noted later that they all made sketches and measured drawings (which in 1925 were still in office scrapbooks), none of these drawings has yet come to light.[6] And perhaps more regrettable, many of the old Colonial buildings that they sketched have disappeared, as McKim feared.

A hint of what they may have learned from this experience can be gauged, perhaps, from two shingled houses that McKim and his two new partners, Mead and Bigelow, designed in 1877–78. These were large, linearly

FIGURE 7. *McKim, Mead & Bigelow. Samuel Gray Ward house, "Oakswood," Lenox, Mass., 1878 (demolished). From C. A. Morely,* Lenox (*East Lee, Mass., 1886).*

FIGURE 8. *McKim, Mead & Bigelow. Mrs. A. C. Alden house, "Fort Hill," Lloyd's Neck, Long Island, N.Y., 1879 (greatly altered). From* American Architect and Building News 6 *(August 30, 1879).*

extended houses that later burned down or suffered extensive alterations. The long expanse of the Samuel Gray Ward house, built in 1877–78 in Lenox, Massachusetts, the summer enclave of Bostonians, is gone, but old photographs show its walls completely sheathed in wooden shingles, and contemporary illustrations show eighteenth-century-style interior paneling (fig. 7).[7] For Fort Hill, built in 1879 for Mrs. A. C. Alden on Lloyd's Neck, Long Island, atop remnants of earthworks from a Revolutionary War fort, McKim used extended horizontal surfaces covered in shingles that pull the design together, as do the matching gables repeated throughout the plan (fig. 8). Although still partially extant, the house has been greatly changed, but the original design is evident in the published elevation.

Although detailed references to specific shingled buildings McKim, White, Bigelow, and Mead saw on their 1877 sketching trip are lacking, it seems likely that these four tourists saw such shingle-clad buildings as the Hoxie house at Sandwich, Cape Cod, Massachusetts, 1637, the Ivory-Board-man house in Saugus, Massachusetts, circa 1725 (demolished circa 1925; fig. 9), and almost certainly the Poore Tavern in Newbury built about 1700 but demolished around 1900. More remote but still a possibility is that they saw the Jethro Coffin house on Nantucket Island, 1686. All of these houses were sheathed entirely in wooden shingles, roofs and walls. One Colonial detail of which the four traveling architects certainly took note was the use of broken glass, embedded in stucco, to form decorative wall patterns. In his reminiscences of the 1877 sketching trip, Mead mentioned only one building specifically, the house of Benjamin Perley Poore near Newburyport, Massachusetts, which contained portions of such stucco work. Poore was an antiquarian who from roughly 1850 to 1880 had acquired portions of many old New England buildings slated for demolition, using them to embellish numerous rooms he added around the seventeenth-century core of his family house.[8]

American Influences

Another influence on the creation of the Shingle Style in the 1880s was the treatment of space in the American houses of the 1870s that Vincent Scully and Henry Russell Hitchcock called the "Stick Style." This term, invented by Scully and then popularized by Hitchcock, described an architecture, largely of summer and country houses, linked to High Victorian Gothic architecture, that featured vigorous external massing determined by relaxed internal room arrangements. Externally these houses were comprised of rather disjointed or gawky assemblages of parts and details, including towers, and they were almost always wrapped at the base with generous porches that reached around two or sometimes three sides of the house. Scully's name derived from the intense woodenness of the stick work in exposed gable trusses of these houses, from framing members defining wall panels, and especially from the exposed basketry of wooden posts and diagonal braces in the encircling porches. Perhaps the best example is the Jacob Cram (later the Sturtevant) house, 1872, in Middletown, Rhode Island, attributed to local builder Dudley Newton (figs. 10, 11). As a comparison of the plan and the exterior reveals, the extreme articulation of each of the faces of the house was determined by the arrangement of the rooms within. From the sweeping porch, the entry vestibule leads to a stair hall around which a parlor, dining room and octagonal library are arranged; to the rear are the kitchen and servants' quarters. There is no constraint on any room to fit a symmetrical or axial alignment; rather, each room is free to shape itself, as it were, around internal requirements. This flexibility of planning was the great gift of High Victorian and Stick Style architects to their successors, who created the Shingle Style.

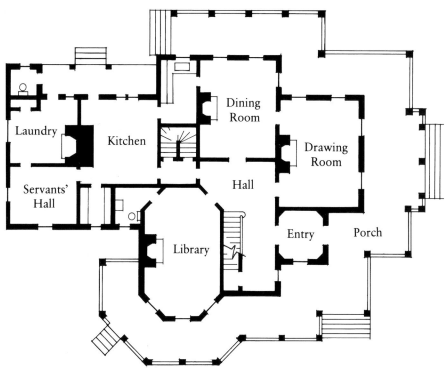

FIGURE 10. *Dudley Newton. Jacob Cram (later the Sturtevant) house, Middletown, R. I., 1872. Photo courtesy of Newport Historical Society.*

FIGURE 11. *Cram-Sturtevant house plan. L. M. Roth, after Scully,* Shingle Style *(New Haven, 1955).*

Japanese Influences

The architects who created the Shingle Style were also greatly influenced by the English and American branches of the Aesthetic Movement. One of the attributes of this movement was the "discovery" of Japanese architecture and Japanese arts and crafts.[9] This intense interest and the resulting *Japonisme* came about, in part, because of the relative flood of trade goods into Europe and America following Admiral Matthew Perry's forcible opening of Japan to foreign trade in 1854. In the world's fair exhibitions that followed, Japan presented pavilions whose traditional architecture and displays of Japanese arts excited great public interest. As European and American businesses began to set up Japanese offices, Westerners began to look more and more keenly at Japanese life and art. One of the centers of this American interest in Asian art (most especially Japanese art) was Boston.

In its drive to modernize, Japan established new universities modeled on Western examples, stressing European languages and the sciences. In 1877, a new Imperial University was created in Tokyo, prompting invitations to many Western scholars to come and teach. One of these professors was Edward S. Morse, of Salem, Massachusetts, who taught zoology in Japan but who soon was consumed with an interest in Japanese art and architecture. He collected voraciously (later his collections passed to the Boston Museum of Fine Arts). Especially alarming to Morse was what he saw as the increasing rejection of certain traditional aspects of Japanese culture, in particular what he foresaw as the eventual disappearance of the traditional Japanese house. His response was to step up his collecting—in this case of information regarding the design, construction, and traditional uses of the parts of the Japanese house. In 1886 his data and interpretation appeared in *Japanese Homes and Their Surroundings*, published by Ticknor and Company of Boston. In addition, for many years Morse had been giving lantern-slide-illustrated lectures on Japan and its traditional architecture. The appearance of Morse's book served to further increase public interest, leading to more invitations to give public lectures farther afield, as he did several times in Chicago in 1890–91.[10]

This influence of Japanese architecture was already quite evident in the United States, especially in early summer houses by McKim, Mead & White. Two examples, regrettably both lost, included the Victor Newcomb Beach house at Elberon, New Jersey, 1880, with its continuous *ramma* rail or beam running around the large living hall, and delicate wood spindle-work screens above the *ramma*, together with the rug pattern that suggested the banded edges of *tatami* mats (figs. 12, 13). The other lost example is the Cyrus H. McCormick summer house, once at Richfield Springs, New York, 1880–82, whose extended porch roofs were supported by dark green painted columns lathe-turned to resemble bamboo (fig. 14). Similar columns were soon used by McKim, Mead & White in their Isaac Bell house, Newport, while various Japanese decorative motifs were used internally.

FIGURE 12. *McKim, Mead & White.*
Victor Newcomb beach house, Elberon,
N. J., 1880 (greatly altered). Interior of living
hall. From G. W. Sheldon, Artistic Houses
(New York, 1885).

FIGURE 13. *Newcomb house, floor plan.*
From M. G. Van Rennselaer, "Recent
Architecture in America: American Country
Dwellings, II," Century Magazine 10
(June 1886): 211.

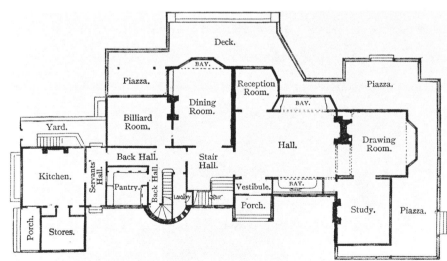

FIGURE 14. *McKim, Mead & White. Cyrus H. McCormick summer house, Richfield Springs, N.Y.,*
1880-82 (demolished). McKim, Mead & White Collection, New York Historical Society.

Medieval Influences

FIGURE 15. *Unidentified photograph of urban house covered in slate, from scrapbook, McKim, Mead & White Collection, Avery Library, Columbia University, New York. (Possibly medieval French.)*

There was at least one further visual source for wooden shingle sheathing, and that was the use of slate in medieval Europe as a sheathing material. Whether this practice had any impact on other architects is not known, but it certainly seems to have affected McKim, Mead & White and Bernard Maybeck. A difficult and heavy material, but one providing extremely long life, slate can be hung on the sides of buildings as well as on roofs if a heavy armature of nailers and supporting rafters has been provided. Holes must be delicately punched in each piece of slate to receive nails or lashing. Evidence for the influence of slate sheathing on McKim, Mead & White is found in some of the photographic scrapbooks assembled by the partners and used as references in later years (and kept today in the Avery Architectural Library of Columbia University). Unfortunately, the snapshots are almost never accompanied by any written caption, so internal evidence alone must provide whatever identification we can make. And, of course, there is no evidence as to when these photographs were made or acquired. Maybeck, too, collected photographs during his years in France, one of which shows medieval slate-covered houses on the Rue de Paris in Vitré, Brittany.[11]

One snapshot, seemingly made or purchased by Stanford White during his trip to Europe in 1878–79 (when he was twenty-five years old) shows the hospital, the Hôtel Dieu, at Beaune with its enormous and steeply sloped roof covered with tiles laying in undulating irregular lines. In letters back to his family, White described Beaune and spoke glowingly of its hospital. As

FIGURE 16. *Unidentified photograph of house covered in slate, from scrapbook, McKim, Mead & White Collection, Avery Library, Columbia University, New York. (Possibly medieval French-Alsatian.)*

for the texture of the hospital roof, he suggested much the same quality in the wavy lines of the slate roof of a building he drew rising over an alley in Bernay. This drawing and others have been published in color by Claire Nicolas White in *Stanford White: Letters to His Family* (New York, 1997). An earlier book of White's drawings was published by his son, Lawrence Grant White, *Sketches and Designs by Stanford White* (New York, 1920), which includes sketches of Lisieux with its picturesque medieval houses, which White also visited. Fresh with the impression of these sights, and with his sketches in hand, White returned to New York at the end of the summer in 1879; within months he was invited to become a partner in a new firm with his friends McKim and Mead. His evident love of the textures possible in slate and tile were almost instantly translated into the early Shingle Style masterworks of the young firm.

The firm's surviving photographs show urban street scenes, with four-story medieval houses, each floor cantilevered out more than the one below, walls, gables, and dormers all covered with slate. In one instance, the slate is cut in patterns running in horizontal bands; along the bottom of the flared wall of the second story, the slate is notched and set in a saw-tooth pattern that stands out against the deep shadow of the beam below (fig. 15). Another shows a winding street descending a hill; in the foreground is a building with a projecting second-story round bay, capped by a tall conical roof. Again, everything—wall and roof—is covered in slate (fig. 16). Here is the model (at least for McKim, Mead & White) of the corner tower with its curved bell-cast roof that appeared so prominently and so frequently in many of the firm's Shingle Style buildings.

An American Architecture of Continuities

Compared to American domestic architecture in the 1870s, the new form of architecture called the Shingle Style stands out because of several unique characteristics. Perhaps because so many of the younger architects then busy inventing the Shingle Style had some measure of formal academic training—many at the Ecole des Beaux Arts in Paris—one defining characteristic is a sense of carefully studied geometry and a sense of an underlying geometric order and relatedness of parts. There are intimations of playful informality and happy accident and coincidence in these structures, but examination of the designs reveals that little is accidental. This relates to what is perhaps the single most important design element: *continuity*—of surface (everything is shingles), line, roof planes, even of space within the building—so that all seems organically connected. Compare this to the disjointed angular nature of so much domestic architecture before this, including the Stick Style and also what was known in the 1870s and 1880s as Eastlake.

This continuity of flowing surface material was introduced in the C. J. Morrill house in Bar Harbor, Maine, by William Ralph Emerson, 1879,

FIGURE 17. *William Ralph Emerson. C. J. Morrill house, Bar Harbor, Maine, 1879. Perspective view. From* American Architect and Building News 5 *(March 22, 1879).*

among the first Shingle Style buildings completed (fig. 17). In perspectives of the house, published in March of that year, Emerson effectively indicates the texture of the shingles, extending from the foundation water table all the way up to the highest roof ridge line.[12] John Calvin Stevens, another of the great innovators of the Shingle Style, was equally skilled in revealing the continuous sweep of shingles in his drawings, as in that of his own house in Portland, Maine, 1884 (fig. 18). Stevens's partner, Albert Winslow Cobb, was equally adept, as is seen in his strongly simple drawing of the firm's house for C. A. Brown in Delano Park near Portland, Maine, dated 1889 (fig. 19). Continuity of surface material and texture encouraged continuity of planes as well. A striking example of the continuity of long roof surfaces is seen in the rendering of the T. R. Glover house of Milton, Massachusetts, 1879, by William Ralph Emerson (fig. 20).

FIGURE 18. *John Calvin Stevens. J. C. Stevens house, Portland, Maine, 1884. Perspective by J. C. Stevens. From J. C. Stevens and A. W. Cobb,* Examples of American Domestic Architecture *(New York, 1889).*

FIGURE 19. *J. C. Stevens and A. W. Cobb. C. A. Brown house, Delano Park, Portland, Maine, 1889. Perspective by A. W. Cobb. From Stevens and Cobb,* Examples of American Domestic Architecture *(New York, 1889).*

This continuity of surface gave rise to the idea of enveloping geometric forms, especially the broad, low-slung triangular gable. This image was first suggested in the broad gable that runs across the front of Richardson's Sherman house, Newport, 1874–75 (fig. 5), but the gable became more emphatic in a house by McKim, Mead & Bigelow (the short-lived predecessor firm to McKim, Mead & White) in a house designed for a good friend, William Dean Howells, in Belmont, Massachusetts, 1877–78.[13] Called Redtop due to

FIGURE 20. *William Ralph Emerson. T. R. Glover house, Milton, Mass., 1879. Perspective by W. R. Emerson. From* American Architect and Building News 6 *(August 2, 1879).*

FIGURE 21. *McKim, Mead & Bigelow. Prescott Hall Butler house, St. James, Long Island, N.Y., 1878-80. Watercolor on paper, 12" x 27". (Photo: L. M. Roth.)*

its sheathing of redwood shingles, the Howells house still stands, but later owners covered the expanse of shingles in the broad gable with stucco. In 1878–80, the firm designed a similar broadly gabled house on Long Island for Prescott Hall Butler. This house still stands as well, although the dramatic triangle is partly obscured today by a dense curtain of trees. The impact of the broad, low-slung triangle that forms the house can best be seen in a watercolor perspective apparently prepared a few years after construction by Stanford White (fig. 21). Other similar triangular house designs followed, including the summer retreat for McCormick at Richfield Springs, New York, and the seaside George V. Cresson house at Narragansett Pier, Rhode Island, 1885. The culmination of these was the firm's elemental triangular gabled house for William G. Low at Bristol, Rhode Island, 1885–87 (fig. 22), now demolished, but rediscovered and made famous by Henry-Russell Hitchcock and Vincent Scully. Every part of the Low house was contained in the enormously extended gable, including the porch at one end. Along the walls, the shingles seamlessly swell out to form the roofs over the bay windows. Other architects of the period also discovered the power of the triangular gable, as did Bruce Price in many of the cottages he designed in the mid-1880s for the romantic retreat of Tuxedo Park. An especially emphatic use of this form is found in the House by the Sea published by John Calvin Stevens in 1889 (fig. 23).

Coupled with this concept of the continuity of surface and form was the continuity of space within the house. As McKim, Mead & White's Newcomb and Bell houses (and many examples by other architects) illustrate, interiors were thought of not as individual rooms in the sense of closed boxes but as spaces that organically opened and flowed into one another. Frank Lloyd Wright would take this concept to its fullest and possibly most eloquent extension, describing the process later in his autobiography by saying that he

FIGURE 22. *McKim, Mead & White. William G. Low, Bristol, R. I., 1885-87 (demolished). (Photo: Historic American Buildings Survey, Library of Congress.)*

FIGURE 23. *John Calvin Stevens. "House by the Sea," 1889. Perspective from Stevens and Cobb,* Examples of American Domestic Architecture *(New York, 1889).*

began to think of the house not as a big box with lots of little boxes squeezed inside but as one continuous space subtly screened into subspaces arranged for separate uses. McKim, Mead & White had already done this in the Newcomb house (fig. 12) by using yawning openings between the living hall and its surrounding spaces, and by opening up the transoms with airy spindle-work grills. In the Bell house they used huge sliding doors so that the idea of the enclosing wall was replaced with that of sliding screens, as in a traditional Japanese house. The huge sliding doors were especially effective in conveying interconnectedness of space, since the opening was far broader than that provided by a simple doorway.

: A HOUSE BY THE SEA :

In Renaissance and Baroque building design, sequences of rooms are arranged *enfilade*, with the doors aligned on one continuous axis. Standing on that axis, the viewer can see the linear succession of doors receding into the distance, hinting at the spaces of those rooms. American architects modified this idea and shifted the axis, so that rooms are placed diagonally in set-back fashion, with the axis through the broad sliding doors running on the diagonal. The result is a more open and dynamic sense of sequential spaces.

Dissemination of the Shingle Style

Clearly something about the freedom of plan and form, and the visual texture of continuous shingles, appealed to a great many architects and their patrons, for the Shingle Style spread quickly across the nation. This came about in part through the publication of designs in both plan books and professional journals, and also through positive critical essays on new suburban and country houses by Mariana Griswold Van Rensselaer that were published in the mid-1880s in *The Century*, a popular literary magazine. In re-examining the Shingle Style in *The Architecture of the American Summer* (1989), Vincent Scully, a native of New Haven, Connecticut, who had grown up among such houes, included examples of shingled houses from magazines published in Chicago and other cities of the West, such as *Inland Architect* and *Western Architect*, which had been left out of his earlier book. Among the buildings added by Scully in 1989 are examples from Detroit (by Mason & Rice), Minneapolis (Cass Gilbert and Leroy Buffington), Cheyenne, Wyoming (William L. Bates), and San Francisco (A. Page Brown). There are also examples built in Florida, Maryland, Virginia, even Ontario and Manitoba, Canada.

Particularly active Shingle Style architects in the West were the Newsom brothers of San Francisco, whose practice extended from Eureka in northern California to Los Angeles. Aside from designing hundreds of houses throughout California (and in other states)—with those of the 1890s being mostly in an elaborate Queen Anne shingled style—the Newsom brothers published at least eleven plan books, five of which featured shingled designs. Most closely analogous to Cobb and Stevens' book of New England designs was the Newsoms' book *Modern Homes of California*, published in San Francisco in 1893, filled with plans for Shingle Style houses bearing names appropriate to the region such as Escondido, Faralone, Pacific Heights, and El Dorado, among many others.[14]

Another reason the Shingle Style spread so far and so quickly was that the large Eastern architectural firms, in particular McKim, Mead & White, trained dozens of younger men who soon moved westward and established themselves in other cities. Graduates of the McKim, Mead & White office are numerous and notable, including Cass Gilbert, who returned to his native St. Paul-Minneapolis, Minnesota, and built numerous shingled houses there, and A. Page Brown, who resettled in San Francisco.

Another McKim, Mead & White graduate was William Whidden, who in 1887 was sent to Portland, Oregon, to finish the hotel begun there by McKim, Mead & White for Henry Villard in 1882. Whidden was almost immediately joined by Ion Lewis, a former associate in the Boston office of Peabody & Stearns, who were masters of the Shingle Style. (Their seaside house for G. Nixon Black, Kragsyde, now destroyed, was one of the great achievements in the Shingle Style). Although Whidden & Lewis built mainly formal, classical urban houses instead of suburban or beach houses, one of their early residences in the fashionable in-town neighborhood of Nob Hill, Portland, was built for Dr. Kenneth A. J. Mackenzie, 1892–94 (fig. 24). In this house the architects combined massive and rough-faced Richardsonian masonry and broad, round arches in the lower elements, but drew on the Shingle Style for the upper walls and roof. Instead of wood shingles, however, they substituted slate in a return to one of the original sources of this idiom. This is one of the rare American houses to employ slate as a wall sheathing material.

Also active in Portland, Oregon, was another transplanted easterner, Ellis F. Lawrence, who had studied at the Massachusetts Institute of Technology and then worked for several New England architects, particularly John Calvin Stevens. In fact, Lawrence had been sent to San Francisco in late March 1906 to supervise a job being started by Stevens. Lawrence had stopped in Portland, Oregon, to visit a friend when the earthquake and fire devastated San Francisco early in April. With his job in San Francisco now gone, Lawrence decided to settle in Portland, and within a year had built for himself a double-shingled house, half for his own family, the other half designed for his mother and sister (fig. 25). About five years later he built a shingled inn at an isolated Pacific beach called Neah-Kah-Nie, frequented

FIGURE 24. *Widden & Lewis. Dr. Kenneth A. J. Mackenzie residence, Portland, Oregon, 1892-94. (Photo: Paul Macapia.)*

by artists, literary figures, and architects. This inn is credited as providing a seminal inspiration for the wooden Northwest regional modern architecture later advanced by John Yeon and Pietro Belluschi in the late 1930s.[15]

In Chicago, the link to eastern shingled architecture was made by Joseph Lyman Silsbee, who moved from Syracuse (and Buffalo) to Chicago about 1882. Perhaps the most persuasive of his shingled houses were those he designed as an ideal middle-class residential enclave for the new northern Chicago suburb of Edgewater in 1886–87, several of which were published in *Inland Architect*.[16] It was in Silsbee's office that young Frank Lloyd Wright had his firsthand introduction to the shingled house. Moreover, when Wright moved to the Adler & Sullivan office and decided to borrow on his future salary to build his own house in Oak Park, he was already familiar with the shingled houses being designed by F. R. Schock in Austin, a suburb immediately to the east of Oak Park (now swallowed up by Chicago). Most of Schock's houses were later replaced with newer or larger residences, including the strongly geometrical and triangular shingled house Schock designed for the town's developer (fig. 26). Gone, too, is Schock's shingled commuter depot in Austin. But the house Schock designed for himself, a fussier, shingled Queen Anne design, survives with only modest changes.

The large and well-known Chicago architectural firms, such as Holabird & Roche and Burnham & Root, also used the Shingle Style in domestic and other buildings, few of which survive, including the commuter station Burnham & Root designed for the suburban community of Buena Park (also today swallowed up by Chicago). Arguably the best of their shingled work was the Lake View Presbyterian Church, 1887–88, now gone, with its commanding tower wrapped with spiraling shingles on the long tapered spire (fig. 27). For the Montezuma Hotel, a large resort built in 1885 on a hillside north of Las Vegas, New Mexico, Burnham & Root used shingles on the upper walls and on the huge corner tower with its soaring conical roof.

FIGURE 26. *F. R. Schock. Austin residence, Austin (Chicago), Illinois, c. 1885-87. Collection of Charles Allgeier proof prints, L. M. Roth.*

The Shingle Style was brought to southern California by a number of Eastern architects. One of the most enigmatic, but masterful of this new idiom, was Ernest Coxhead, an English-born-and-trained architect who (with his brother) set out for the United States in 1886 and headed directly to Los Angeles to design Episcopal churches there.[17] Four years later, the brothers resettled in San Francisco, where Ernest designed churches as well as residences. The grandest example of Coxhead's ecclesiastical work was his huge Church of St. John the Evangelist, San Francisco, 1890–91, regrettably dynamited to create a firebreak after the earthquake in April 1906 (fig. 28). Perhaps an even more masterful example of his fluid use of shingling is the more intimate Chapel of St. John, Monterey, California, 1891. The shingles are treated like a mass of water swirling and curling over rocks on the shore, sweeping around corners and flowing over stone buttresses and arched openings. How Coxhead managed to so thoroughly assimilate the American Shingle Style while passing through the East remains uncertain, but master it he did. Many of his smaller shingled churches survive, and one of those in the best condition is his St. John's Church in Petaluma, California, 1890.

One of the catalytic figures in bringing the Shingle Style to San Francisco was the peripatetic Willis J. Polk. Born in Jacksonville, Illinois, and raised in St. Louis, he traveled and worked for a number of architects, including Van Brunt & Howe in Kansas City and the young A. Page Brown in New York.[18] He relocated with Brown to San Francisco in 1889. By 1902 he was in Chicago working for Daniel Burnham, then was sent back to San Francisco to head Burnham's branch office there for several years. A highly gifted artist

FIGURE 27. *Burnham & Root. Lake View Presbyterian Church, Lakeview (Chicago), Illinois, 1887-88. Courtesy of the Burnham Library, Art Institute of Chicago.*

and delineator, Polk produced superb drawings for his employers, and he also published several designs of his own, including a rectilinear shingled house designed for Kittie Beardsley at Plymouth, Massachusetts, in 1889 (just before he left for San Francisco), and a shingled house for F. W. Rosenthal in Alameda, California, 1890, just after he arrived on the West Coast. In 1892 he designed a remarkable shingled double house for Mrs. Virgil Williams and his own family on an extraordinarily steep hillside site in San Francisco. Other shingled Bay Area houses by Polk followed in the 1890s.

Polk's colleague and friend Arthur Page Brown was born in Ellisburg, New York, studied briefly at Cornell University, and then spent about four years in the office of McKim, Mead & White.[19] In 1884, with the patronage

FIGURE 28. *Ernest Coxhead. Church of St. John the Evangelist, San Francisco, 1890-91 (demolished). Photo courtesy of Richard Longstreth.*

of Mrs. Cyrus McCormick, Brown started an independent practice in New York City, in the same building as McKim, Mead & White, at 57 Broadway. Then, with promising commissions from the Crocker family, he relocated to San Francisco in 1889, taking with him his chief draftsman, A. C. Schweinfurth, and soon hiring Polk as well. In San Francisco, Brown was an interpreter of the urbane classicism of McKim, Mead & White, but his earliest work, before and after the move to San Francisco, drew on his mentors' shingled work as well. Willis Polk prepared ink perspectives for several of Brown's shingled designs published in mid-1889, including the Crocker Old People's Home, which drew on McKim, Mead & White's design for the Northern Pacific Railroad hospital in Brainerd, Minnesota.[20]

At the same time that A. Page Brown and Polk were working so close to McKim, Mead & White in New York, Bernard Maybeck was working for Carrère & Hastings in New York, themselves recent graduates from the McKim, Mead & White office. Maybeck had grown up in New York City, the son of a transplanted German artist and cabinetmaker.[21] While Brown was serving his apprentice years in the office of McKim, Mead & White, Maybeck traveled to Europe and studied several years at the Ecole des Beaux Arts, returning to New York in 1886 and entering the new office of Carrère & Hastings. Maybeck was soon persuaded by Brown and Polk to begin an independent career in San Francisco, but before he left New York a hint of his stylistic inclinations was given in a shingled house for E. H. Johnson in Greenwich, Connecticut, published in *American Architect and Building News* in August 1888. Although technically a project of Carrère & Hastings (and signed by Hastings), it is believed to have been largely a Maybeck design, and incorporates the creative informality that would later characterize Maybeck's work in California.

Once in California, Maybeck would, in turn, influence Julia Morgan, who would, in her own way, embrace the use of shingles for informal churches, clubs, and houses. Hence in the intertwined careers and work of Polk, Coxhead, Maybeck, Schweinfurth, Morgan, and others the use of shingles as an expression of bohemian creativity and artistic freedom would be introduced to San Francisco and around the Bay Area, establishing a regional tradition that would flourish for several generations.

As the illustrations in *American Summer* reveal, there was considerable interest in shingled architecture in Los Angeles and Pasadena, California. Several shingled houses in Pasadena by the Greene brothers in the Arts and Crafts tradition are included in Scully's book, but there are also several other shingled houses, approaching the eastern definition of the Shingle Style, by architects Hudson & Munsel, H. Ridgeway, Blick & Moore, and F. L. Roehrig. For Los Angeles there was a large shingled hotel designed by Franz E. Zerrahn, as well as residences by W. Redmire Ray, Locke & Munsell, Train & Williams, Hunt & Eager. Another Hunt—Myron Hunt—arrived from Chicago in 1903, forming a partnership with Elmer Grey. Myron Hunt had been a colleague of Frank Lloyd Wright in the Steinway Building in Chicago and was, like Wright, a charter member of Chicago's Arts and Crafts Society in 1897. Myron Hunt lived in suburban Evanston, where he designed several carefully modeled shingled houses. Not long after arriving in Los Angeles, however, Hunt abandoned shingled architecture in favor of an idiom he deemed more in character with the history and climate of California.

The Shingle Style did not come to an end at any fixed moment, but as economic and social conditions changed it enjoyed a long twilight well into the twentieth century, continuing to shape a few special houses while undergoing a transformation from within. A spectacular example of a late Shingle Style house is Tenacre in Southampton, New York, designed in 1922 for the publisher Joseph P. Knapp (fig. 29). The Knapp house is surprising on several levels, and it is hard to say which is more surprising: the architect or the very late date. Tenacre is very much in the tradition of the expansive, generously scaled shingled summer house, with its two large twin flanking gables. But it was not the kind of severe, grandiose house usually associated with Pope at this date, when he had already begun designing monumental classical buildings such as the grandly domed train station for Richmond, Virginia; the latter hints at the Jefferson Memorial and the National Gallery

FIGURE 29. *John Russell Pope. Joseph P. Knapp house, "Tenacre," Southampton, Long Island, N.Y., 1922. From* American Architect–The Architectural Review 122 *(September 13, 1922).*

of Art, which he would design in the 1930s. Pope was a former protege of McKim, and his penchant for classicism clearly comes from this source. But Tenacre shows the more informal insights he may also have garnered from his contacts with McKim, Mead & White, as well as from his experience as a designer in the office of Bruce Price.

With the introduction of the personal income tax, rising wages, and decreasing numbers of live-in servants, the commodious shingled summer house of the 1880s and 1890s slowly became a thing of the past. Yet the sources of inspiration for the American Shingle Style—the English Arts and Crafts movement, which stressed visible construction and unconcealed expression of building materials, the inspiration from Japanese arts and crafts, and William Morris's echoing injunction to make a complete architecture that was within the reach of ordinary working people—continued to exert a force that merged aspects of the Shingle Style into the growing bungalow and Craftsman house movement in the teens and twenties. The use of banded shingle siding, the horizontality, the sense of protective enclosure in dark natural-wood paneled interiors, the emphasis on a comfortable and uncomplicated way of living in the bungalow and Craftsman house, all attest to the continuing influence of the Shingle Style into the 1930s, when the Depression interrupted private building for almost everyone.

The Shingle Style Discovered (or Rediscovered)

In the 1880s and 1890s, this American shingled residential idiom was called Queen Anne, borrowing a term invented in England to describe the free classical asymmetrical composition devised in the 1860s by architects like Nesfield, Shaw, Bodley, and others, in red brick structures (with white painted wood trim in windows and doors), embellished with classical details from the time of Sir Christopher Wren.[22] (The term, in fact, had virtually nothing to do with British architecture actually built during the reign of Queen Anne, 1702 to 1714.) In the United States, the term Queen Anne was freely and broadly applied to residential architecture that was asymmetrical in composition, skirted with wraparound porches, punctuated by corner towers capped with conical or bell-cast roofs, and usually employing shingles as a partial wall covering as well as roofing material. It was an enormously popular house style.

By 1930, however, the Queen Anne style had fallen out of critical and public favor. Two decades later, at its nadir of appreciation as modernism seized the public imagination, Queen Anne domestic architecture became the subject of scholarly study by Vincent Scully. After earning his bachelor's degree at Yale University and serving as a Marine during World War II, Scully returned to Yale for his doctorate in art history, focusing on the wood-framed houses of the last third of the nineteenth century. The way for such a re-examination had been blazed a decade before in the relatively sympathetic treatment given to Henry Hobson Richardson's houses in Henry-Russell Hitchcock's ground-breaking study, for Hitchcock argued that Richardson was a prophet of modernism.

Scully, however, would examine the entire sweep of American framed residential design, from A. J. Downing in the 1850s to the emergence of Frank Lloyd Wright in the 1890s. What Scully undertook, furthermore, was to place Wright in a continuum of American residential design, showing how he adapted, abstracted, and clarified what had been done before, how Wright analyzed all this in the context of the mid-American landscape and then developed the horizontally emphatic Prairie House. Scully's dissertation was called "The Cottage Style," but from that study emerged the two terms generally describing the early and later phases of this framed residential architecture that have established themselves in the lexicon of modern art and architectural studies: the Stick Style and the Shingle Style.

In 1955 the larger part of Scully's dissertation appeared as *The Shingle Style: Architectural Theory and Design from Richardson to the Origins of Wright,* published by Yale University Press. It was an immediate critical success. When it eventually went out of print, so intense and persistent was the demand that in 1971 Yale University Press reissued the book, adding the first part of Scully's dissertation and using the revised title *The Shingle Style and the Stick Style.*

In Scully's usage, especially, Shingle Style came to mean that subset of Queen Anne houses in which a uniform covering of shingles was used, and which celebrated the informality of American domestic life with sweeping continuities of surface, line, and space. For him the luminous achievement of the Shingle Style was the then relatively unknown William G. Low house by McKim, Mead & White, reduced to elemental form and stripped of extraneous detail. To Scully (and to Hitchcock before him) it seemed to point the way to future modernism. Curiously, the Low house had been unpublished and unheralded in its own time, and was very little known except in Bristol, Rhode Island. Through Scully's efforts, the Low house became famous, now known by every architecture student, and through his rediscovery of this house and all those that led up to it, a recognizable form of uniquely American architecture was given its own name.

Almost immediately after finishing his doctorate, Scully was appointed to teach at Yale, developing a lecture style that in content and delivery captivated architectural students. Visitors to his lectures were common and included visiting architects teaching in the Yale architecture school. In the 1950s, 1960s, and 1970s, Yale became one of the world centers of higher education for architectural historians, and by the early 1970s Scully was a celebrity.[23] In February 1980, Scully was the subject of a profile in *The New Yorker,* and when he retired in 1991 his final lecture was covered on the front page of *The New York Times.* His students by now held positions of leadership, both in architectural criticism (Paul Goldberger) and in architectural design (Robert A. M. Stern). Moreover, many visiting architects teaching at Yale

FIGURE 30. *Robert Venturi and John Rauch. Wislocki and Trubeck houses, Nantucket Island, Mass., 1971-72.*

during the 1960s and 1970s had came under the pervasive influence of Scully and the Shingle Style, so that aspects of the Shingle Style began subtly to shape or influence contemporary design, either in the flexible handling of mass and space or in the use of shingles as a sheathing material. The list of architects is impressive: Charles Moore, Frank Israel, Charles Gwathmey, Romaldo Giurgola, Turner Brooks, Giovanni Pasanella, and most especially Robert Venturi, as seen best in his pair of cottages, the Wislocki and Trubeck houses, on Nantucket, 1971–72 (fig. 30). Although it was not yet evident in 1973, Robert Stern would take the Shingle Style he learned from Scully and make of it a major part of his version of Postmodernism. Scully himself commented on this remarkable influence of history on contemporary design in a lecture he gave at Columbia University in September of 1973, published in 1974 as the slender book *The Shingle Style Today: or, The Historian's Revenge.*

Robert Stern's rise in the worlds of architectural design, criticism, and education was meteoric. While continuing as a prolific designer, in the late 1970s he was also appointed head of the newly formed Temple Hoyne Buell Center for the Study of American Architecture at Columbia University. There, he initiated a program to publish source documents in American architecture. Among these was a revisiting by Scully of the Shingle Style (then heavily influencing Stern's own work). The result was called *The Architecture of the American Summer: The Flowering of the Shingle Style*, with an introduction by Scully followed by 180 pages of large reproductions of shingled houses, hotels, clubs, and churches, as they appeared in professional and popular journals from 1875 to 1922. As Scully readily admitted in his introductory comments, the focus of his studies in the 1940s and 1950s had been largely limited to the Northeast. In this new assemblage of published images of

Shingle Style buildings, this narrow focus was broadened to include churches, hotels, club houses, and residences from coast to coast, from Canada to Florida. In the 1980s and 1990s elements of the Shingle Style have been embraced again for new residential work as a kind of nostalgic recreation by architects like Robert A. M. Stern and his many followers, and in more abstracted, contemporary ways by architects like Jaquelin Robertson, Turner Brooks, Daniel Solomon, and Jeremy Kotas. The Shingle Style continues to be a potent force in the creation of American architecture.

WATTS SHERMAN HOUSE

Newport, Rhode Island, 1874–76

A huge gable and strong horizontals mark this first monument of the emerging Shingle Style.

OPPOSITE: *The upper stair hall is lit by an expanse of windows. The glass panes are painted with stylized sunflowers, an emblem of the Aesthetic Movement.*

To summer in the height of fashion at the end of the nineteenth century one spent July and August in Newport, Rhode Island. There, in the 1860s, 1870s, and 1880s, the wealthy and the ambitious clustered along Bellevue Avenue and along the eastern shore of Aquidneck Island in enormous houses they referred to as "cottages."

Newport, established in 1639, had grown to become a major shipping port by the mid-eighteenth century, with finely wrought houses built for wealthy merchants and ship captains. This was what could be called, architecturally, Newport's First Golden Age, the period of builder Richard Munday and gentleman-amateur architect Peter Harrison, and the period during which some of the most exquisite Georgian structures to be found anywhere along the Atlantic coast were built.

By the end of the Revolutionary War, however, other ports had drawn away Newport's lucrative traffic and the town began a period of genteel somnambulance. Before the Revolution, West Indian plantation owners had customarily journeyed to Newport during the summer months, and they were followed in the nineteenth century by plantation owners from Georgia and South Carolina who began to make Newport their summer place of residence.

At the conclusion of the Civil War, a new generation of wealth began to expand the summer colony, and the building boom of grand summer cottages started. Old-time Newport residents called it "the New York invasion." In the mid-1870s this cottage-building provided the impetus for the beginning of the Shingle Style, and its first monument was the William Watts Sherman house, designed by Henry Hobson Richardson.

After designing the Watts Sherman house, Richardson burst upon the American architectural scene with a broad popularity no other architect before him had enjoyed. The buildings that caused this were his Trinity Church, Boston (1874–77), and the Watts Sherman house. Both had a pleasing exuberance of detail controlled by powerful elemental massing in deep earth-related colors.

Richardson's clients were William Watts Sherman and his bride, Annie Derby Rogers Wetmore. William was a lawyer from New York City, but Annie was from the prominent Rhode Island Wetmore family. Her brother,

The south side of the house shows the influence of the picturesque English houses of Richard Norman Shaw.

OPPOSITE: *In 1881 Stanford White made the drawing room into a library. Now bereft of books, it remains a dazzling synthesis of classical, oriental, and Arts & Crafts motifs.*

George P. Wetmore, was governor of Rhode Island and then a U.S. senator. Their father, William Shepard Wetmore, made a fortune in the China trade and was one of the mid-century cottage builders. He engaged Seth Bradford in 1851 to design the proud vertical mansarded granite pile called Château-sur-Mer, which was sensitively enlarged by Richard Morris Hunt two decades later and made even more vertical, with taller mansard roofs. In 1871, soon after his marriage, William W. Sherman purchased a parcel of land next to Château-sur-Mer from his father-in-law and hired Richardson in 1874 to design his summer house.

Where Château-sur-Mer lifts itself up grandly, Richardson's compact Watts Sherman house hugs the earth, especially when seen from the front, where extended overhangs, a band of banked windows, and the cantilevered upper gable all create long horizontal lines. The rear facade, with its more numerous gables and projecting window bays, shows how Richardson was looking at the English country houses of Richard Norman Shaw of a decade earlier, houses Richardson had seen in England and which had been published in elegantly crisp photoengravings of Shaw's own perspectives. Like Shaw, Richardson used a mix of stone and brick masonry, with half-timbered-looking walls, but Richardson made a decisive change from Shaw's wall covering, employing wooden shingles instead of Shaw's heavy red clay tiles. This opened up possibilities for variations in texture and surface, with the shingles cut and nailed on in different patterns, which Richardson exploited, especially in the upper gables.

When first built, this and other shingled houses had roofs covered likewise with wooden shingles, creating a continuity of color and texture now diminished wherever roofs have been recovered with asphalt composition roofing. Additionally, the Watts Sherman house had emphatic horizontal lines in the roof surfaces, created by long raised lips in the roof shingles; the dark shadow lines created by these selected rows of shingles coincided with intersection points of dormers or roof ridges.

The raised rows of roof shingles are also now gone, and the domestic character of the house has suffered as a result of its being converted into a college dormitory, with requisite security lights and cameras, fluorescent fixtures, sheet vinyl floors, and other incongruous appurtenances. Still important for their original spatial qualities and detailing are the huge central living hall, incorporating a generous staircase of multiple landings and a broad, hooded fireplace, as well as a library refurbished by Stanford White with dark forest-green paneling and delicate gold-leaf ornamental accents.

The potential for plastic manipulation of shingled surfaces, only hinted at in the Watts Sherman house, was not lost on Richardson's able assistant, Stanford White, then twenty-one years of age. Within five years of starting work on the Watts Sherman house, White would join his friends McKim and Mead to form a new partnership that would push the possibilities of shingling nearly to the limit.

NEWPORT CASINO

Newport, Rhode Island, 1879–81

The flattened gables facing Bellevue Avenue only hint at the delights of the Casino grounds.

OPPOSITE: *A latticed porch overlooks the lawns and tennis courts of America's first country club.*

There is in the Shingle Style work of McKim, Mead & White a playful and effortless charm that sums up the character of young Stanford White. Even late in his life, shadowed by financial problems and personal demons, White was famous for his boundless enthusiasm and hearty embrace of life. His youthful insouciant charm is evident in every line of the Newport Casino.

The Casino came about as the result of a cheeky bet made by James Gordon Bennett. Mercurial and impulsive, Bennett was publisher of *The New York Herald*, famed for presenting lurid stories and scandal but also famous for Bennett's sending Henry Stanley off to Africa in search of Doctor David Livingstone. Bennett was also a sportsman. In Newport, he introduced a number of activities that quickly became part of the new social order: polo, steeplechase, squash, golf, ocean yacht racing, and seashore clambakes.

Bennett had brought over an entire British polo team to play against his own, and heading the British team was a Captain Candy. While riding past the Newport Reading Room one day, a member of the club (according to some, Bennett himself) dared Candy to ride his horse into the club, which Candy did. His action so outraged members of the club that they voted to rescind Candy's honorary membership. This, in turn, so infuriated Candy's sponsor, Bennett, that he resolved to quit the Reading Room and build his own club, which would become the Newport Casino. Bennett reckoned, correctly, that the smart set would follow him in joining the new facility.

Bennett acquired the large parcel of land next to the Travers Block, a Stick Style mixed-use building at the corner of Bellevue Avenue and Bath Road. His new clubhouse was to contain club rooms, guest rooms for bachelors, lounges, a restaurant, theater, and tennis courts on the lawns behind the building. Ground-floor shops would flank a central arched entry passageway leading back to the heart of the Casino. From the moment it opened in 1881, the Casino was the center of social life among the summer colony. Virtually every kind of sport played on grass was supported: croquet and cricket, as well as the more vigorous lawn tennis (then a novelty from England) and American baseball. The Casino Theater saw a period of lavish entertainment, with elaborate dress balls held there, as well as concerts, readings, and amateur and professional theatrical productions. The Casino became the focal

The arched entranceway.

OPPOSITE: *The broad "piazza" around the center courtyard.*

point of lawn tennis, with national tournaments held there in 1881. When tennis became a professional sport rather than a social pastime, the national tournaments were shifted to Forest Hills Gardens.

White and his partners provided all the right spaces, especially the shaded broad verandahs, called piazzas, for casual strolling. The piazzas are screened from the lawns by decorative spindle basketry, partly inspired by the spindle screens in Japanese houses. The screens around the piazzas create dappled patterns of light while also artfully framing views toward the main club house and out across the tennis lawns.

Aside from the brick ground floor walls and piers, the entire building is covered with shingles. Along Bellevue Avenue, the facade is symmetrical, with a broad central gable over the recessed loggia flanked by telescoped matching gables. Inside the Casino, however, the symmetry is relaxed, with the facade facing the inner court composed of dormers, varied windows, and a bold tower in a balanced but asymmetrical arrangement. The tower tapers inward toward the top, and the cap was given a delicate bell-shaped curve. The large clock at the top served a most useful function, for aside from being a picturesque accent, the clock face could be read over the roof of the piazzas, a convenience for the players who did not wish to wear their pocket watches while on the courts.

The airy delicacy of the piazza screens was echoed in the filament-thin furnishings of the lounges, as evident in period photographs. In the now-abandoned theater one can see a similar lightness of touch. The rectangular theater room is lined with balconies behind broad elliptical arcades on either side. The wooden trim of the walls—the slender pilasters and arches—is classical in detail, with the frieze of the entablature that runs around the room at the ceiling filled with playful images of seahorses and scallop shells.

The Casino Theater provided entertainments for Newport society.

OPPOSITE: *Light-hearted classical ornament for a summer theater.*

The spandrel walls between the arches are covered with a basket-weave pattern in the plaster. The balcony railings incorporate slender turned balusters (based on eighteenth-century Colonial models) interspersed with square grids that take their inspiration from Japanese *ramma* grills. Here is a tour-de-force of delicate ornament, picked out in accents of gold leaf.

The Casino's adroit melding of sharp bilateral symmetries, with the casual ease of fusing delicate Colonial and classical details with spindle-work derived from Japanese sources, was without precedent in the United States. It perfectly suited the mix of formal elegance and informal spontaneity that characterized the summer colony of the early 1880s. The studied order of McKim and the carefree invention of White were perfect complements, and the Casino was the beginning of their Shingle Style.

KINGSCOTE
DINING ROOM
Newport, Rhode Island, 1880–81

The shingled addition harmonizes with the earlier Gothic Revival house.

OPPOSITE: *In the dining room, Stanford White melded lavish materials with such diverse architectural influences as colonial paneling, Japanese spatial articulation, and Moorish and Art Nouveau detailing.*

OVERLEAF: *A mahogany spool-and-spindle screen at the entrance to the dining room, which also served as a ballroom.*

By the mid-nineteenth century, some of the summer visitors had tired of the crowds of the Newport hotels and decided to acquire land and build their own residences. One of the first was the Savannah plantation owner George Noble Jones. In 1839 Jones hired architect Richard Upjohn, who had just finished a house in Maine for the Gardiner family into which Jones had recently married. Jones desired a Newport summer cottage on property he acquired on Bellevue Avenue, then on the outskirts of Newport. Bellevue Avenue would become the most desired location for the wave of cottages built after the Civil War.

The Maine Gardiner house, admired by Jones, was a battlemented Gothic structure. For the Jones cottage Upjohn devised a more picturesque and irregular Gothic design, a rambling one-story affair, with a prominent two-story polygonal bay flanking the recessed entrance. Clad in flush-fitted boards, with dramatically steep-pitched roof dormers trimmed with heavy bargeboards and brackets, it was painted gray to suggest stone construction. It established a new note of romantic fantasy for summer architecture in Newport, for, as Upjohn wrote, the Gothic style was capable of greater variety of form and construction than any other style.

With the outbreak of the Civil War, Jones had to abandon his Gothic cottage in Newport. In 1863 the house was sold to Jones's Newport neighbor and friend, William Henry King, who had made a fortune in the China trade. King, who never married, suffered a mental collapse only a few years after acquiring the house and was committed to a sanatorium; his business affairs were then administered by members of the family. A nephew, David King, Jr., who also had become wealthy through the China trade, was appointed guardian of the house in 1875 and married that same year. His bride, Elle Louisa Rives of Virginia, was entranced by the house, and the couple soon undertook alterations to bring it up to date.

The younger King was a close associate of Stanford White, then starting work on the Newport Casino directly across Bellevue Avenue from the King house. In 1880–81, White and his firm were engaged to make dramatic changes to the King house, now called Kingscote (Kings Cottage). White left the front portion of Upjohn's house untouched but moved the kitchen and

The built-in sideboard in the dining room. After the 1876 Centennial, spinning wheels became popular domestic props among the nostalgic well-to-do.

OPPOSITE: *A sinuous handle graces a sideboard door.*

service wing back forty feet, inserting in the opened space a large new dining room on the ground floor, with bedrooms and baths above that rise to an octagonal shingled tower that echoes Upjohn's polygonal bay on the opposite side of the house.

The dining room was oversized, furnished for family dining but also capable of being converted into a small ballroom for entertaining. The entire room is paneled in mahogany up to a height of seven feet, with a cornice of delicate classical moldings forming an uninterrupted band running around the room and continuing across the windows. This molding, inspired by the *kamoi* beam in Japanese house construction, also forms the upper part of a spindle lattice screen that frames the entrance to the dining room, creating that teasing, splintered view of the space beyond that White also exploited around the verandahs of the casino. On one side of the room, the paneling is transformed into a built-in, recessed sideboard, with spindle doors, and capped by three scallop-shell arcades inspired by Colonial detailing in Newport architecture and by eighteenth-century Newport furniture. The frieze above the mahogany is covered with strips of amber-colored cork laid in a basketwork pattern repeated in the ceiling panels. This was a fortunate material choice, for not only does the color of the cork complement other materials in the room, its sound absorptive qualities make the large room more acoustically intimate.

The warm hues of the parquet floor, mahogany wall paneling, and cork-sheathed walls and ceiling are just the beginning of the sumptuous coloration that White delighted in. The large fireplace is set in a wall sheathed with panels of golden Siena marble, with a narrow bronze band around the hearth mouth, so that the room-encircling cornice becomes the fireplace mantle as well. Above the cornice mantle, the lines of the fireplace wall continue upward, but here the wall is covered with small variegated moss-green Tiffany tiles. The delicate grid of the green tiles is picked up by the equally small squares of opalescent green and blue Tiffany glass that form a window wall on either side of the fireplace. Because the glass tiles and the fireplace marble surround are set flush with one another, there is a sense of sheer flatness that denies the presence of the wall, and yet the hearth itself projects deeply behind this implied plane. The expanse of translucent glass and the shimmering, shifting colors create a sense of intimate enclosure and privacy, while opening up the wall to the sun at the same time.

The Kingscote dining room is one of the great surviving Shingle Style interiors. Its air of relaxed elegance is all the more remarkable compared to the heavy-handed Beaux Arts interiors that would proliferate in Newport in a few years. White showed, playfully, how sumptuousness could be achieved without pomp and pretense.

ISAAC BELL HOUSE

Newport, Rhode Island, 1881–83

Of all the summer houses that ushered in the Shingle Style, the finest surviving example of this free-spirited and festive architecture is the Isaac Bell house, built for James Gordon Bennett's sister and her husband. A decade later the summer cottages of the very wealthy would become massive piles of stone inspired by European Renaissance and Baroque palaces, but the Bell house was a light confection of frame construction, of carefully balanced triangular masses and a tower, all sheathed from roof ridge to the brick first floor wall by wooden shingles, and wrapped by an encircling open porch. Thus the Bell house marks a critical point in the architectural history of Newport's summer colony, and indeed in the history of American architecture as a whole.

RIGHT: *Porches overlook the lawn and fashionable Bellevue Avenue.*

OPPOSITE: *The Bell house shows McKim, Mead & White's skill at resolving asymmetrical elements into an ordered but dynamic whole.*

A window beside the front door.

The house was created for Isaac Bell, Jr., and his bride, Jeanette Bennett, whose energetic brother had commissioned the Newport Casino just two years earlier. Bell, a highly successful cotton broker, had inherited considerable wealth from his father but also built up his own business so that he was able to retire in 1877 at the age of thirty-one. The next year he married Miss Bennett at her brother's Newport estate, and soon after they began planning their Newport house. The Bells moved into the house in 1883, but because of Bell's service as the U.S. ambassador to the Netherlands during the Cleveland administration, he and the family were able to spend only three summers in the house before his death in 1889 at age forty-one.

The free composition of the house, with its expansive gables recalling nearby eighteenth-century houses, was markedly original. The novelty of the design gave rise to some confusion in contemporary descriptions. The Newport *Mercury* in October 1881 described it as a "Queen Anne villa," showing that this term was already being used in the 1880s. But George W. Sheldon, in his elegant volume *Artistic Country Seats* (New York, 1886), described it as being in a "modernized Colonial style," and he was correct, too, for certain elements were inspired by old Newport models. On the exterior, one decidedly old-fashioned feature was the use of small panes of glass in the windows, recalling eighteenth-century glazing, instead of large sheets of modern glass.

What makes the exterior so visually appealing is its sense of order achieved through the balance of related but dissimilar elements. The two broad gables do not repeat the same window pattern, and the solid round tower on the south side is echoed in the rounded voids of the superimposed porches on the east side. Although the upper wall surfaces are organized into continuous horizontal bands, the shingled surfaces have varying textures. Some sections are square cut, other portions are laid with undulating curves, and still other areas are covered in rounded fish-scale shingles. The subtlest of all these details is the single course of notched shingles at the line of the middle sash bar of the double-hung windows that introduces another faint horizontal line to the composition.

As in the Casino, an encircling porch or piazza is a principal external feature. The entry steps, on the south side, are covered with a rounded extension of the piazza roof, supported by large brackets carved in a dolphin motif, which was also used at the Casino. The piazza melds into the tower that rises through the second floor and is capped by a bell-shaped roof. The porch columns are tapered and turned to resemble lengths of bamboo.

A notable feature of the house is its sense of openness, emphasized by the horizontal flow of interior and exterior spaces. The reception room, drawing room, dining room, and study are arranged around a spacious living and stair hall, nearly 30 feet by 24 feet, the biggest room of all.

The rooms are connected by broad openings containing huge sliding doors. The opening from living hall to drawing room stretches to sixteen

OPPOSITE: *The reception room leads through broad doors to the drawing room, with its floor-length windows that open onto the porches.*

The exquisite interior architecture of Mrs. Bell's room.

feet; its four doors roll on overhead rails. Oversized exposed hardware was created to indicate the doors' mobility, and although the surfaces of the metal straps have Japanese-inspired decorative details, the idea of these large exposed wheels and rails comes from the doors of New England barns.

Particularly notable is the spindle-work and paneling around the great hall fireplace, made up of panels from a Breton bed alcove. The spindle-work is set against back plates of polished tin to reflect splinters of light. Flanking the fireplace are windows made up of small panes of beveled glass that splash rainbows of afternoon sun across the dark expanse of the hall. One can see the same care in merging small-scale detail in Mrs. Bell's bedroom, on the second floor, where the plaster basketwork in the upper frieze (like that in the Casino theater) becomes an open grid that runs in front of the larger grid of the corner window bay. This bedroom paneling reprises much of White's scheme for the Watts Sherman library. The paneling around the fireplace was originally painted cream and pink, with detail picked out in silver-leaf—a flamboyantly feminine world.

The interiors of the Bell house, especially, show McKim, Mead & White's connection to the Aesthetic movement, which attempted to infuse beauty into all areas of design and living.

The chief proselytizer of the Aesthetic movement, poet and aesthete Oscar Wilde, arrived in Newport during the summer of 1883, giving a celebrated lecture in the Casino theater entitled "The Beautiful." Wilde exhorted his audience to seek this delicate and exquisite fusion of all forms of design, creating an architecture of "Sweetness and Light," a phrase made famous by Matthew Arnold in his book *Culture and Anarchy* (1869).

The Aesthetic movement drew special attention to traditional Japanese art and furniture. Strict symmetry was shunned in favor of near-symmetrical compositions achieved by the balance of dissimilar elements. This sensibility is what informs the design of the Bell house, with its paneling drawing from the eighteenth century; its spindle-work and continuous plate-rail *kamoi* molding inspired by Japanese sources; its panels of woven rattan; its "bamboo" porch columns.

Of all the extant shingled houses by McKim, Mead & White, the Bell house best illustrates the characteristic elements of their Shingle Style—the openness of space, expansive porches, delicate, small-scale decorative details, and references to Japanese art and architecture and the English Arts and Crafts movement. After a long period of neglect and decay, the house is being restored by the Preservation Society of Newport County.

SAGAMORE HILL

Oyster Bay, Long Island, 1883

The house was placed on a hilltop with sweeping views of Oyster Bay and Long Island Sound.

OPPOSITE: *Lamb & Rich designed the living hall as a shadowy refuge from the out-of-doors.*

I wonder if you will ever know how I love Sagamore Hill.
—THEODORE ROOSEVELT

These were very nearly the last words Theodore Roosevelt said to his beloved wife Edith. At age sixty-one and in ill health, he spent his days reading and writing in his bed at Sagamore Hill as Edith sat with him. The two had raised their six children here and the children of Roosevelt's brother grew up nearby. As many as sixteen youngsters gamboled about Sagamore Hill in the 1890s. Many afternoons, Theodore and Edith spent hours alone together, hiking or riding in the earlier years, quietly reading in the later years. As the afternoon of January 5, 1919, lengthened, Edith rose to step from the room, and Teddy spoke of their house to her. Later that night, he passed away in his sleep.

Theodore Roosevelt was born into an old New York family, descendants of early Dutch colonists. The family was comfortably well-off; they were not among the *nouveaux riches* of the city. The Roosevelts lived in a townhouse on East Twentieth Street in Manhattan and summered in rented houses around Oyster Bay on the north shore of Long Island. Young Teddy was encouraged to swim and ride and hike with his many cousins, as a way of strengthening his body and fighting his severe asthma. He became especially fond of climbing wooded Sagamore Hill—named after the Indian chief Sagamore Mohannis, who had sold this land to early Dutch settlers—to declaim poetry into the wind at the top of his voice. Later, as a young man, newly graduated from Harvard, Teddy resolved to build a house atop Sagamore Hill.

In 1880 Roosevelt married Alice Hathaway Lee, and in January 1884 they began to make plans for a new house at Oyster Bay. It was to be a year-round house for a large family, with twenty-three rooms, eight fireplaces, and two hot-air furnaces. But in February Alice died after giving birth to a daughter; later that same day, Teddy's mother also died. Nearly destroyed by grief, Teddy went to North Dakota to lose himself in a rugged life of cattle ranching. But before going West in March, he signed a contract for building the house atop Sagamore Hill. The architects selected were Lamb & Rich of New York City. Eventually Roosevelt returned to New York and married again, to Edith Kermit Carow. Although Sagamore Hill had been completed by mid-1885 at

Late afternoon in Mrs. Roosevelt's parlor.

a cost of $16,795, Roosevelt and his bride did not move into the house until the winter of 1887. (Roosevelt's infant daughter, named Alice after her dead mother, was already living in the house, cared for by her Aunt Anna.)

Hardly had the newlyweds settled in when Roosevelt learned that a bitter winter on the high plains had wiped out his cattle. Desperate at the loss of his only source of steady income, Roosevelt considered selling Sagamore Hill. Later, in equally desperate financial straits, he would again consider selling Sagamore Hill. But each time he resisted. The house would become his beloved refuge from the pressures of political life, the place he yearned to be most. Roosevelt's daughter Alice, who later married Nicholas Longworth (Speaker of the House of Representatives from 1925–31), wrote of Sagamore Hill in her reminiscences: "we resented the winter months we had to spend away from it in Washington and New York and counted the days until it was time to go home again." For the family, Sagamore Hill was indeed "home."

Little is known about how Lamb & Rich came to be selected by Theodore and Alice Roosevelt as their architects. Hugo Lamb and Charles Alonzo Rich had formed a partnership in 1882, which continued until Lamb's death in 1903. Rich, who had worked in Boston for William Ralph Emerson, was a gifted designer, a prolific watercolorist, and a world-traveler. He formed enduring friendships with many clients, including Roosevelt. Lamb & Rich designed apartment houses, churches, theaters, and important complexes for Barnard, Smith, Dartmouth, and Colgate Colleges. But today they are best remembered for their vigorous shingled work of the 1880s. Perhaps their best residence of this period was Sunset Hall (1883), the S. P. Hinckley house at Lawrence, Long Island. It was published in G. W. Sheldon's showpiece book *Artistic Country Seats* (New York, 1886–87). The linear extension of the plan, the opening of the rooms to broad porches, the many gables filled with patterned shingles, all would reappear in the design of Sagamore Hill.

A Lamb & Rich sketch for Sagamore Hill published in *American Architect and Building News* in 1893 shows extremely steep gables, almost a burlesque of medieval houses. As built, the gables are at a more serene slope of 45 degrees. The walls of the first floor are of red brick, accented at window openings by sunflower medallions of cut brick. Roosevelt recalled in a letter of 1915, "I did not know enough to be sure what I wished in outside matters, but I had perfectly definite views [regarding] what I wished in inside matters, what I desired to live in and with." These included a "library with a shallow bay opening south; the parlor or drawing room occupying all the west end of the lower floor; as broad a hall as our space would permit; big fireplaces for logs; on the top floor a gun room occupying the western end so that . . . it looks out over the sound and bay." Lamb & Rich devised the plan exactly

On the piazza.

OPPOSITE: *The library fireplace.*

as Roosevelt instructed, with a broad living hall extending through the house, opening to the parlor, library, and dining room through broad sliding doors. Just as Roosevelt requested, wrapped around the house was "a big piazza, very broad at the N.W. corner where we could sit in rocking chairs and look at the sunset." Lamb & Rich's color scheme for the house was as bright and cheerful as the owner himself, with the brick painted a rich orange, the shingles mustard yellow, and the trim red and green. Such vivid polychromatic treatments were characteristic of the shingle houses of Lamb & Rich in the early 1880s.

The library, festooned with trophies from Roosevelt's big game hunts, served as his summer office when he was president. The parlor was Edith's feminine preserve, off-limits to the rambunctious boys. The porches were favored by all the family; they were expanded in 1905, and large awnings were installed to further shade them from the southwest. Alice later recalled her father's fondness for them: "The evenings in summer and early autumn are very lovely at Sagamore. Father used to dress in time to get downstairs a little before dinner, and go out on the piazza to watch the sun set behind the distant blue lines of Connecticut across Long Island Sound. I think he cared most for the blazoned autumn sunsets that left a long clear afterglow. He would stand on the piazza and look and look; start to go in and go back for another look."

STONEHURST

Waltham, Massachusetts, 1883–86

The balcony outside Mrs. Paine's room.

OPPOSITE: *Within a vigorous composition, a Palladian window pays homage to the ancestors of Richardson's clients.*

Although McKim, Mead & White were among the principle innovators of the Shingle Style, many of their landmark shingled houses have been lost. Their Newcomb house was radically rebuilt after a disastrous fire; the McCormick and Low houses were both demolished.

Henry Hobson Richardson has been more fortunate. His Watts Sherman house and Stoughton house both survive, although somewhat modified. Richardson's last country house, Stonehurst, designed for Robert Treat Paine (1835–1910), survives virtually intact, combining Richardson's bold inventive forms with interiors that represent perhaps the pinnacle achievement of Shingle Style architects and craftsmen. Externally, it represents a fusion of the massive glacial boulder construction of Richardson's celebrated Ames Gate Lodge at North Easton, Massachusetts, 1880–81, with the all-shingle construction of his Reverend Percy Brown house of Marion, Massachusetts, 1881–82, his Stoughton house in Cambridge, Massachusetts, 1882–83, and his compact cottage for the Ames estate gardener, also in North Easton, 1884–85.

Robert Treat Paine came from a distinguished Boston Brahmin family, descended from the Robert Treat Paine who had signed the Declaration of Independence. Paine had studied at Harvard, graduating just as Richardson began his studies there. In 1859 he was admitted to the bar and practiced for the next thirteen years. Through astute investments in railroads and mining property, Paine became wealthy, and he retired from his law practice in 1872 to devote his energies to philanthropic endeavors. In 1863 Paine had married Lydia Lyman, whose ancestors had hired Samuel McIntire in 1793 to build a fine classical house in Waltham in the middle of a large estate that came to be called the Vale. Upon Lydia's marriage, her father gave the couple a parcel of land in the Vale. They soon built a sedate mansarded Second Empire Baroque house designed by Boston architect Gridley J. F. Bryant.

After ending his legal practice, Paine served on many public boards and committees, one of which was the building committee formed to rebuild Trinity Episcopal Church after the disastrous Boston fire of 1872. In this way Paine became acquainted with Richardson, who was awarded the commission to build the new church on Copley Square in Boston's Back Bay. The rector of Trinity, the Reverend Phillips Brooks, gave great credit to Paine for the success of Trinity's building campaign. Margaret Henderson Floyd,

ABOVE: *Boulders from the estate were used in the house and its retaining walls.*

OPPOSITE: *A nook in the summer parlor.*

OVERLEAF: *The vast living hall.*

SECOND OVERLEAF: *The living hall fireplace and window seat.*

Richardson's recent biographer, quotes Brooks' warm thanks in a letter to Paine, saying that Paine saved the church "from doing things on a small scale, and kept us large . . . we owe it to you that Trinity Church is big and dignified, and not a little thing in a side street, which one must hunt to find." Paine must have remembered that building campaign with satisfaction, for later he kept a perspective of the church over the fireplace in his study at Stonehurst.

In 1882 Lydia Paine's father died and she inherited her portion of his estate. The Paine family, now with five children, was somewhat cramped in the original Bryant house. The inheritance would make possible a grand expansion. In the autumn of 1883, Paine asked Richardson to expand his house in the Vale. He also approached Frederick Law Olmsted to devise a new placement for the house and to draw up landscape plans. By the next year, 1884, Olmsted and Richardson had settled on a new location for the house, on a high ridge with a fine view to the southeast. The Bryant house was to be moved to this new setting and have a new southerly section added by Richardson, making the original house the service wing. Realizing that the enlarged house would now require earth embankments on either side for stabilization, Olmsted suggested using the soil removed in excavating the new basement to create these terraces north and south of the house, and he designed long serpentine retaining walls of large rough boulders. On the south front, these retaining walls were to curve back on themselves in steps

Mr. Paine's study.

descending to the lawn. (In construction Paine modified these as earth ramps rather than stairs.)

Richardson designed what was essentially an entire new house, linked to the old house at its northwest edge. The corners of the new section were marked by massive round stone towers, built of irregular granite boulders, with windows framed with rough slabs of red sandstone, so that the squat roundness of Richardson's towers echoes the sweeping curves of Olmsted's terrace wall. An early perspective sketch of the house shows what appear to be shingled lower walls, but as built the first-floor walls are also of granite boulders, as are the supporting piers of the loggia on the south front. The eastern end wall has a yawning semicircular arch of these boulders that seems to push the broadly flared shingled wall of the second floor up and out of the way. As Floyd observes, the east elevation is quintessential Richardson, flaunting asymmetry, marrying massive elements to delicate ones, juxtaposing Colonial classicism with the almost Chinese flare of the lower roof line. The earlier perspective proposed tall shingled conical roofs over the towers, but in construction these were eliminated in favor of flat platforms that provided sweeping views of the surrounding countryside.

Richardson also deviated from his early design in lowering what was to have been a tall hipped roof to make it a low-pitched gable roof, with a broad bank of small-paned windows lighting the attic. All the windows have

the thinnest of framing architraves—mere lines—in contrast to the heavily framed and decidedly neo-Colonial Palladian window in the center of the end wall. With this device, Richardson pays homage not only to Lydia's nearby family home (which has a similar window by McIntire) but also to Paine's illustrious eighteenth-century ancestor.

As in Richardson's other houses, the plan moves around a huge central living hall, running the width of the house, with a fireplace and a grand stair—the epitome of such Shingle Style staircases—that rises in broad, easy stages and includes built-in seating, small-scale paneling, and layer upon layer of beautifully turned delicate balusters. The ceiling is deeply coffered with exposed beams, and, to ensure the grandest sweep of space, uninterupted by any support posts, the massive beam that runs the width of the house past the stair is supported at its center by an iron tie-rod that drops down from the roof structure. Clearly visible on the bottom of the beam is the wrought-iron plate and nut. Richardson's loving attention to detail is demonstrated in the rounded inglenook off the northeast corner of the great hall, with a built-in window seat that curls into a volute at its end, the armrest supported by thin spindles that echo those used in the stair railing. In the great hall, and also in the summer parlor at the southeast corner of the house, the transoms are opened up with some of the finest spindle-work of the period, another example of Japanese inspiration in the Shingle Style. If spindle-work details here and there are reminiscent of McKim, Mead & White, this is more than a coincidence, for recent research has revealed that Richardson owned a large collection of photographs of the Newport Casino, and the influence of his former assistants appears frequently in the interiors of Stonehurst.

Unlike the even more massive granite Glessner house in Chicago, which has long enjoyed appreciation and acclaim thanks to frequent publication (although it was finished posthumously by Shepley, Rutan and Coolidge, Richardson's successor firm), the Paine house has remained a sleeping beauty, despite the attention that Richardson was able to devote to it.

OPPOSITE: *The southwest tower contains the stair landing (top) and a low-ceilinged parlor under the landing (below).*

NAUMKEAG

Stockbridge, Massachusetts, 1884–87

The Shingle Style was an architecture of exuberance free of gauche display, of whimsy without affectation. These qualities are found at every turn at Naumkeag, the summer house of New York lawyer Joseph Hodges Choate. Some of New York's elite, like Choate, eschewed the excesses of Newport society for the quiet Brahmin atmosphere of the Berkshire Hills, around the village of Stockbridge, in western Massachusetts. There, as access to the outside world became easier with the arrival of the railroads in mid-century, farms depleted by years of hard agriculture were sold to wealthy New Yorkers and Bostonians who set out to build commodious though unpretentious summer places, landscaping the abandoned fields to form picturesque glades and forests.

One of the most notable of this new gentry was Choate, a trial lawyer of considerable "finesse" (as Brendan Gill put it). He won over judges and juries

RIGHT: *The hillside site affords broad vistas of the gardens and the Berkshires.*

OPPOSITE: *An arch in the driveway wall looks into the gardens beyond.*

FIRST OVERLEAF: *The stair hall.*
SECOND OVERLEAF: *The dining room.*

A guest bedroom.

with the practiced wit and charm that made him a popular after-dinner speaker. Active in the Republican Party, he was appointed Ambassador to the Court of St. James by President McKinley, serving from 1899 to 1905. His successful legal challenges of the personal income tax assured, for a time, the viability of the spacious country house as an architectural idiom.

Choate, born in Salem of a family established in Massachusetts since the seventeenth century, graduated from Harvard College (1852) and Harvard Law School (1854), beginning his law practice in Boston. In 1855 he relocated to New York, forming a partnership with William Evarts and Charles Butler. Butler, fourteen years Choate's senior, had a home in Stockbridge, to which he retired in 1881. Others of Choate's acquaintance, such as the Sedgewicks of Boston, also owned houses in the Berkshires, so in 1884 when Choate decided to build a summer house he knew where he wanted it and who the architect would be.

In 1881 Choate had served as the official speaker for the dedication of the Admiral David Farragut Memorial in New York's Madison Square, and was impressed with the young architect, Stanford White, designer of the unusual base for Saint-Gauden's bronze figure. Only two years in the firm of McKim, Mead & White, White had already designed a number of summer houses, including, in 1879, a Long Island house for Prescott Hall Butler, a young member of Choate's law firm. Butler would soon become White's brother-in-law, marrying the oldest daughter of Judge J. Lawrence Smith of Smithtown, Long Island; White wed the judge's youngest daughter in 1884.

Choate's expansive character was matched by that of Stanford White. The house White designed for Choate on Prospect Hill outside Stockbridge embodies the delight in the picturesque and unusual they shared. Choate called the house Naumkeag, the native name for his birthplace, Salem, said to mean "Place of Rest" or "Haven of Comfort." Set in an estate of forty-nine acres, this house was to be Choate's retreat from the pressures of New York.

The original landscape design was developed by Frederick Law Olmsted, who placed the house lower on the slope of Prospect Hill, but McKim is said to have argued in favor of placement nearer the top of the hill because of the better views. Eventually Olmsted was excused and replaced with landscape designer Nathan Barrett.

Naumkeag is a house that defies easy stylistic categorization. It includes bits of all that White had observed in his studies and travels, including the massive and rough-faced masonry of his mentor, H. H. Richardson, the round towers with their diagonal brick diaper patterns of Normandy, which he had just visited, and details from Colonial architecture, which he had studied in 1877 on the walking trip with his future partners. White employed a wall treatment he saw in several old houses along the New England coast: a coating of thick plaster into which pieces of broken bottles and glass slag had been inserted, forming sparkling decorative patterns. Hence one finds in Naumkeag recollections of all of these experiences, in a relaxed L-shaped plan, terminating in round Norman towers at the far corners. In the mortar between the bricks of the east wall are hundreds of pieces of glass, so that the morning sun generates sparks of light across the surface. One tower encloses

The landscape designer Fletcher Steele used this room during extended visits to Naumkeag to develop the gardens.

"The Blue Steps"
designed by Fletcher Steele.

OPPOSITE: *The childhood bedroom of the Choates' daughter Mabel, with its sleeping porch and hammock.*

a bay in the parlor, while the other houses Choate's personal study. The main block of the house, bisected by a generous hall, recalls the double-pile plan of eighteenth-century Georgian houses. A local newspaper commentator described Naumkeag in 1886 as "somewhat in the Old English Style."

Details and finishes vary from crisp attenuated classical forms—recalling eighteenth-century sources, as in the interior paneling, stair balusters, and fireplace surrounds—to the rough and rustic, as in the peeled pole joists over the entry porch and the rugged fieldstone of the tower bases. Everywhere one finds great attention to detail, as in the curving and shaping of the shingles as they turn to meet the slope of the gable roof, or the sunburst and interlace patterns carved into the fireplace mantelpieces.

Naumkeag passed to the Choate's daughter, Mabel, who kept the house largely unchanged in memory of her parents. She devoted her energies to the grounds, working with noted landscape designer Fletcher Steele to make several pleasantly eccentric gardens. When she died in 1958, Mabel Choate bequeathed Naumkeag and its contents to the Trustees of Reservations, who continue to make the house and its gardens open to the public.

CHARLES LANG FREER HOUSE

Detroit, Michigan, 1890

The exterior suavely plays asymmetrical elements against a symmetrical outline.

The Shingle Style was an attempt to make the private house a work of art possessing an artistic integration equal to the paintings of the Old Masters. If any one house embodied this desire it was the Charles Lang Freer residence in Detroit.

Freer, a diminutive man, dapper in dress and polite in manner, made his fortune organizing and running a company that manufactured railroad freight cars. He used his considerable wealth collecting the most refined contemporary art works; he also amassed one of the world's most important collections of Asian art.

Born in 1854 in Kingston, New York, to a family of high ideals but limited means, Freer went to work as a clerk at age fourteen. Adept at figures, he came to the attention of Frank J. Hecker, then in charge of a local railroad, and Freer soon joined Hecker as accountant and paymaster. In 1876 Hecker relocated to Indiana to supervise the amalgamation of two railroads, and Freer became treasurer of the new company. Aline Saarinen, in an essay on Freer as a collector, noted that "all his life he thought double-entry bookkeeping one of the most beautiful things in the world." In 1885, when their Indiana railroad was acquired by the Wabash railroad, Hecker and Freer moved to Detroit to start a company building freight cars. In 1892 they acquired a competing company, monopolizing the industry in the Midwest. The company so prospered that in 1899, at age forty-five, Freer retired from business to devote himself to his avocation, now become a consuming passion—the collecting of Asian art.

While in the early stages of his rise to wealth, Freer began to educate himself in the visual arts, beginning by collecting prints, first Dutch and then the delicate atmospheric works of Whistler, whom he met in 1888 and who was thereafter one of his closest friends. Freer gradually purchased editions of every etching Whistler produced. Soon Freer was collecting Whistler's paintings as well, and expanding his interests to include the Asian arts Whistler so admired. He also became fond of the work of several American painters, especially the quiet landscapes of Dwight Tryon, the mysterious, atmospheric images of American women by Thomas Dewing, and the glowing virginal female images of Abbott Thayer.

Carved cherubs and grapevine sconces further the Italian fantasy.

Inevitably, the rented rooms Freer occupied became inadequate to house his collection, so he began to seek an architect who could design a setting ideal for his works of art. As he traveled on railroad business, he looked at new houses, learning about their architects. In 1887 Freer and Hecker purchased adjoining lots on Detroit's Ferry Avenue, a fashionable area at the edge of the city, but not too far from their plant. In 1890 Hecker built an imposing, châteauesque house, designed by Louis Kamper, formerly an assistant in the office of McKim, Mead & White. Freer had no desire for such a pompous display. He knew Stanford White and admired the elegantly severe frames White designed for canvases by Whistler and Dewing, but Freer may have worried rightly about White producing an overpowering setting for his paintings. About 1888, Freer saw the house of Charles Potter at Germantown, just outside Philadelphia, and admired its subtle massing and restrained detail, all wrapped in shingles on the upper wall and roof surfaces. Its architect was Wilson Eyre of Philadelphia, then reaching his peak as a designer of Shingle Style houses in the city's fashionable main line suburbs. Eyre was hired in 1890 to design Freer's house on Ferry Avenue. He was also willing to work closely with Freer, so that the house would be a collaborative effort of client and architect. Freer wished to ensure that each detail contributed to the overall aesthetic effect. The house was to be modest, since Freer lived alone and would entertain only a handful of guests at any time.

The plan that Eyre devised was focused on a large central room divided into two areas by a central fireplace mass—two hearths back to back. On the street side was a spacious entry hall, and on the other side was a spacious oak-paneled stair hall rising through two stories. In a possible reference to Japanese screen work, Eyre used carved panels suggesting interwoven strips of wood in place of traditional balustrades. To the left of the entry hall was the library, with several window seats, one nestled next to a small angled fireplace. To the right of the entry hall were the parlor and the dining room. Off the library was a covered semicircular porch, and on the other side of the house was a second covered porch that extended around to the front of the house as an open terrace to the front entry. The lower level of the house was built of a deep brown-to-purple stone from quarries near Kingston, New York, while the upper area was sheathed entirely in cypress shingles flowing across the surface, over and around windows.

Freer corresponded with the Boston decorative firm of A. H. Davenport, ordering the furnishings he needed. He also desired a particular color stain for the oak paneling in the hall, and experimented with applying vinegar to iron to create rust, which when mixed with water achieved the right color. Meanwhile Freer's artist friends were eager to create works to grace the house. Dwight Tryon painted four panels representing the seasons, two seascapes, and a landscape called Dawn that was hung over the fireplace in the entry hall. The proportions of each were determined by the architecture and complemented the horizontal lines of the rooms. All the frames were designed by Stanford White. Tryon worked with New York decorator W. C. LeBrocq to develop squares of Dutch metal, a kind of muted imitation gold leaf, stippled with patches of brown and blue, applied in squares to the walls. For the parlor Dewing created a portrait of his daughter, in a narrow frame designed by White. The largest of the works by the young American painters friendly with Freer was *A Virgin*, by Abbott Handerson Thayer, which was placed at the foot of the stairs.

Few of the Asian pieces were on permanent display. Following a custom of the Orient, Freer had his guests sit in the alcove at the front of the hall, while his servant would bring works out one at a time and place them for study in the light that flooded through the south-facing bay window. Certain that art could be properly appreciated only in natural light, Freer saw to it that when the sun set, the viewing would end.

From the beginning, Freer had never planned on his house becoming a public museum. He made other arrangements, donating his collection to the nation, and engaging Charles Adams Platt to design a sedate classical building to house it next to the Smithsonian Institution on the Washington Mall.

Today the paintings have been relocated to Washington and Freer's house seems somewhat barren. For many years it has housed the Merrill-Palmer Institute. Perhaps the removal of the art allows one to see the architectural elements more clearly. One can still sense the muted vibrancy perceived by architecture historian W. Hawkins Ferry, who wrote that the Freer residence "displayed a sensitivity to line and texture that marked it as one of the most distinguished houses of its period in Detroit."

The library inglenook.

SHELBURNE FARMS

Shelburne, Vermont, 1885–1902

During the decades following the Civil War, there emerged several prestigious watering spots where captains of industry and their wives could pursue their displays of leisure in the summer months. For those desiring sea breezes, there was Elberon, New Jersey, which was soon eclipsed by more fashionable Newport, Rhode Island, and exclusive Bar Harbor, Maine. For those favoring horses, there was venerable Saratoga, New York. There were the rustic camps nestled among the lakes in the high Adirondacks. And there were also the more reserved, Brahmin intellectual pleasures of the Berkshire Hills in western Massachusetts. Yet there were a few individualists who sought out their own retreats, where their souls found personal solace. Such were Dr. William S. and Lila Vanderbilt Webb.

William Seward Webb—the son of James Watson Webb, the fiery owner and editor of the New York *Courier and Enquirer*—was privately tutored, spent a few years at military school, and then finished his college studies at Columbia College (now Columbia University). He then decided on a career in medicine, graduating from the College of Physicians and Surgeons in New York City in 1875, pursuing further studies in Vienna. But on Webb's return to New York, medicine may have proved insufficiently stimulating, for in 1879 he embarked on a career on Wall Street. Within a few years Webb had come to the attention of William Henry Vanderbilt, owner of a vast

RIGHT: *The Farm Barn is the first building glimpsed from the curving drive into Shelburne Farms.*

OPPOSITE: *A tack room in the Breeding Barn.*

network of railroads, becoming an agent for Vanderbilt and proving his worth by rescuing the moribund Wagner Palace Car Company (a few years later he would be named its president). Webb also came to know the Vanderbilt family well, and married the younger daughter, Lila, in 1881.

It was on a scouting mission for Vanderbilt, checking out a small Vermont railroad for purchase and expansion, that Webb was sent north to the dappled hills around Burlington. He fell in love with the landscape, as did his wife. They decided to build a relatively modest summer cottage, called Oakledge. There they could enjoy the view over what is often described as the most beautiful lake in America, toward the distant purple peaks of the Adirondack wilderness to the west. Behind them, to the east, rose the Green Mountains of Vermont, dominated by Mount Mansfield. At first the Webbs wintered in New York, but increasingly Oakledge became their permanent home, and they began to envision a far larger estate there. His private railroad car would enable Webb to make easy overnight trips to New York to attend to business.

When Lila's father died, she received $10 million from his estate of over $200 million. During 1885–86 they began to purchase the separate farms on Shelburne Point jutting out into Lake Champlain south of Burlington, soon assembling thirteen parcels into an expanse of three thousand acres. To plan the landscape, the Webbs consulted with Frederick Law Olmsted, and for the management of the woodlands and farms they conferred with Olmsted's protege Gifford Pinchot. Gently sinuous gravel and macadamized roads were laid through the parklike landscape and a high knoll was selected as the site of a new residence. Besides the house, there was to be an experimental farm, particularly for the breeding of horses. Webb thought too much inbreeding had occurred in the region, producing a skinny horse that could hardly pull a plow. He undertook to breed a handsome general-purpose

A horse stall in the Breeding Barn.

OPPOSITE: *The horses lived in airy neo-Renaissance splendor.*

OVERLEAF: *The vast interior of the Breeding Barn was used for polo during bad weather.*

hackney that could work in the fields but also elegantly draw a carriage and passengers to town.

The house was designed by the Philadelphia architect Robert Henderson Robertson, who had made a reputation designing railroad stations, including depots for the Hudson Valley Railroad, and perhaps in this way came to Dr. Webb's attention. A native of Philadelphia, Robertson had studied in Scotland and at Rutgers University and then entered the architectural office of Henry Sims in Philadelphia. His years there, 1869–72, were followed by brief periods working for George B. Post and then, successively, for the brothers Edward T. Potter and William A. Potter. Robertson's own style had moved from the colorful and angular High Victorian Gothic of the Potters to the more rounded and integrated style of Henry Hobson Richardson, which Robertson was using in the Mott Haven station for the New York Central Railroad in 1885 and 1886, when he was approached by Dr. Webb. During 1886 Robertson produced a series of designs for Webb that included a huge Classical Revival mansion and a large stone castle, until he was reined in and produced a comparatively modest shingled cottage design.

Construction began in the fall of 1887 and the house was finished the next year. Two stories high, it was a linear arrangement of rooms stretching 128 feet long by 50 feet wide, softened by round corner towers and a porte-cochère. On the ground floor were the dining room, breakfast room, library, drawing room, an office for Dr. Webb, and a kitchen with larders. On the second floor were several bedrooms, dressing rooms, two nurseries, a smaller day nursery, the nurse's room, guest bedrooms, and linen closets. Rooms for the servants were tucked under the broad roof.

Business reverses (including a legal battle with the Pullman Palace Car Company, which resulted in Pullman taking over the Wagner Company) persuaded the Webbs to make Shelburne Farms their permanent residence. In early 1895 Robertson began work on expanding the house. The main house was enlarged, the roof raised to create a larger second floor, and the manorial image enhanced by a proliferation of tall clustered and paneled chimneys and half-timbered gables, which recalled the English Jacobean country houses that had been a source of the Queen Anne style. A large wing was added, so that the total number of rooms reached 100 and the plan changed into that of a Y; a broad semicircular covered verandah rounded the end of the new wing, matching the original verandah at the other end of the house. Finished in 1899, this was the largest house in all of Vermont. Perhaps the Vermont winters played havoc with the original shingled wall surfaces, but whatever the reason the facades were covered with brick in 1903.

Beyond the house were gardens, with greenhouses providing three acres of heated protection. The gardens were the province of Lila, who came increasingly under the influence of English garden designer Gertrude Jekyll. She gradually replaced the more formal Italian-inspired plantings and layout with Jekyll's relaxed and more naturalistic forms and indigenous materials.

ABOVE: *The Webb residence at Shelburne Farms grew from a shingled "cottage" into the largest house in Vermont.*

OPPOSITE: *The staircase within the original cottage.*

The north garden centered on a witty sculpture of a nymph astride a fish, given to the Webbs by Stanford White.

The relaxed and informal house, which encouraged an equally informal way of living at Shelburne Farms, stood in marked contrast to the ice-cold classical McKim, Mead & White mansion of Lila's brother Frederick Vanderbilt at Hyde Park, New York. And it is far different from the enormous French Renaissance château, Biltmore, that her brother George had built by architect Richard Morris Hunt in the mountains of North Carolina.

About a quarter-mile south of the house was a large coach barn designed by Robertson, then demolished and replaced with a larger version also by Robertson in 1901–2. The larger coach house, measuring 190 feet by 160 feet, surrounds a central court measuring 100 by 75 feet and dominated by a clock tower.

The outbuildings began with the large Farm Barn, built in 1888–90 from plans by Robertson, and distinguished by a pair of half-timbered gables in the middle of its shingled roof. This building, the center of Dr. Webb's model farm, rises the equivalent of five stories, and the courtyard enclosed by the

ABOVE: *Dr. Webb's room. His valet's quarters were just up the spiral stairs.*

OPPOSITE: *The library fireplace.*

barn and wings is itself nearly two acres. The Farm Barn originally housed shops for carpenters and blacksmiths, staff facilities, and stables for eighty teams of draft mules and horses. Like all of Robertson's other barns on the estate, it is sheathed in shingles, and it originally had a shingled roof.

Even bigger was the Breeding Barn, designed by Robertson and built in 1891. It measures 418 feet in length and is 107 feet wide, enclosing one vast room two stories high, its span made possible by the delicate iron roof trusses. One long gable roof covers the entire expanse, broken up by Robertson with a symmetrical array of dormers. The interior was the largest clear-span space in the United States. During the glory days of Shelburne Farms, it was used for polo matches during bad weather. At one end were thirty-two box stalls, with twelve larger stalls at the opposite end. Across from the Breeding Barn and built at the same time is the Dairy Barn, also by Robertson, which stretches 270 by 45 feet.

The introduction of the individual income tax in 1916 and the Great Depression brought hard years for Shelburne Farms. The proliferation of private automobiles and the introduction of gasoline-powered tractors made Dr. Webb's horses unnecessary. The stables fell into disrepair. The gardens reverted to lawns. The estate passed to the Webbs' children and grandchildren, who lived in houses on other parts of the estate or in other cities. The big house remained closed almost year-round.

But in the early 1970s, the younger generation of the family discovered anew the treasure created by William and Lila; they jointly resolved not to subdivide the land but to make the estate work in some new way. They set up Shelburne Farms Resources, a nonprofit organization with cultural and educational divisions, to manage the woodlands, run a summer camp for disadvantaged children, and establish a farmers' market and a craft cooperative. In 1974 the music room in the main house became the venue for a Mozart Festival, which proved hugely popular, so much so that younger members of

BELOW: *The library retains the Webbs' books and their "fainting sofa."*

The library porch.

the family became the staff of the house, opening up rooms and boarding visitors to the festival as well as the musicians.

During 1983–87, the Big House underwent extensive repair and subtle renovation, opening as a luxurious inn and restaurant. Now a working farm of 1,400 acres and a nonprofit educational institution, Shelburne Farms has become a place where past and future overlap, where visitors can bask in nostalgia for a time of comfortable leisure while pursuing study. Fourth-generation family member Alec Webb wrote "It is ironic that the revitalization of the property to serve a future-oriented mission has given us a new appreciation of its past."

HOTEL DEL CORONADO

Coronado, California, 1886–88

The buildings front a broad beach and the Pacific Ocean.

OPPOSITE: *The hotel is built around a garden courtyard.*

Although the Shingle Style is customarily associated with seaside summer cottages, it was also used by many of its earliest eastern innovators for summer resort hotels as well. In 1875–77 McKim and Mead had each designed shingled hotels before they formed their famous partnership with Stanford White, and that firm itself designed a large shingled hotel for Lewis Roberts in 1882 while also designing and building a large shingled hospital for employees of the Northern Pacific Railroad at Brainerd, Minnesota.

None of these designs was published but other shingled hotel projects were, including Bruce Price's West End Hotel, Bar Harbor, Maine (1879), Franz E. Zerrahn's proposed hotel for Los Angeles, California (1884), Burnham & Root's sprawling Montezuma Hotel, Las Vegas, New Mexico (1885), Clarence Luce's hotel for Cushing's Island, Maine (1887), a hotel at Little Falls, Minnesota, by Cass Gilbert (1888), an enormous hotel proposed for Kingsville, Ontario, Canada, by Detroit architects Mason & Rice (1889), the Pontefract Inn, designed by Hoppin, Read & Hoppin (1890), a proposal for a hotel on Lake George, New York, by Frank T. Cornell (1891), a large house-like inn at Ridgefield, Connecticut, by William A. Bates (1891), and the Red Swan Inn, Warwick, New York, by E. G. W. Dietrich (1904).

This long list suggests the popularity of informal shingled architecture for such retreats. Some were projects only, and many that were built would succumb to fire. In any case, the communal character of hotel design at the time would doom those structures that did survive, as more private pursuits, and private bathrooms, became the requisite norms of twentieth-century resorts. Today, near San Diego, California, one great shingled hotel of the Gilded Age survives. Even at the end of the twentieth century it retains an atmosphere appropriate to its alluring name: the Hotel del Coronado.

San Diego, like Pasadena outside Los Angeles, acquired a reputation in the 1880s and 1890s as a place of ideal weather, therapeutic and invigorating. As the railroads established direct links to eastern and Midwest cities, these small communities flourished as vacationing and wintering places, and as places to regain one's nervous equilibrium and health. San Diego grew up around the original Spanish mission community established in 1769, on arable land next to the San Diego River. Nearby to the north were tidal flats and lagoons that became Mission Bay, while just over two miles to the south was the broad

The outer balconies originally served as open-air corridors. They are now partitioned into private sitting areas.

OPPOSITE: *The repetition of balconies and windows creates some surprisingly modern-looking architecture.*

OVERLEAF: *The billowing volume of the dining room, and its magnificent coffered ceiling.*

crescent-shaped San Diego Bay, protected from the Pacific by a long sand bar. Up until 1885 this bar was barren except for chaparral, rabbits, and quail. Two new San Diego residents, Elisha Babcock and H. L. Story, who had come to San Diego for their health, often rowed over to the peninsula to hunt for rabbits. In 1885, convinced that the boom was just beginning, they purchased the entire peninsula, drew in other partners, and incorporated the Coronado Beach Company to subdivide and develop a new community called Coronado, including the construction of "the largest hotel in the world . . . too gorgeous to be true," as their prospectus promised.

To design their hotel, Babcock and Story hired James Reid, who was then working as draftsman for the Evansville & Terre Haute Railroad in Indiana. Reid had been born in New Brunswick, Canada, had studied at McGill University in Montreal, and then obtained his architectural training at MIT, followed by a period at the Ecole des Beaux Arts in Paris. After finishing the hotel, he would move to San Francisco, where his brother Merritt, in the meantime, had set up an architectural office (they were later joined by the third brother, Watson). The architectural firm of Reid & Reid quickly became one of the most important in that city and designed major buildings up and down the Pacific coast through the first decades of the twentieth century, including the lavish Fairmont Hotel on San Francisco's Nob Hill.

When Reid arrived in Coronado, he was instructed by Babcock (as he recounted in a memoir of 1938) that the proposed four-story hotel was to be built around a rectangular court, 150 by 250 feet, containing "a garden of tropical trees, shrubs, and flowers. . . . Balconies should look down on this court from every story. From the south end the foyer should open to Glorietta Bay with verandas. . . . On the ocean corner there should be a pavilion tower, and northward along the ocean, a colonnade, terraced in grass to the beach. The dining wing should project at an angle from the southeast corner of the court and be almost detached, to give full value to the view of the ocean, bay, and city."

With lumber and skilled labor unavailable in San Diego, arrangements were made to ship green, rough-sawn lumber from the north, and unskilled Chinese laborers were contracted. A planing mill, an iron works for hardware, and a brick kiln were built on the site. Reid started building at the less complex northern side so that the workmen would develop skills as they approached the more complicated lounges, lobby, rotunda, and dining room on the south side. Newly developed incandescent lighting, just perfected by Thomas Edison, was installed, along with back-up gas fixtures, which never had to be used.

Contributing to a sense of shared community, all circulation originally was by means of the exterior balconies, around the courtyard as well as on the sides facing the ocean. Only later were interior corridors created and the balconies converted into private extensions of each room.

Given the early date of its construction on the west coast, the distance from the centers of Shingle Style development and the young age (thirty-five years) of architect James Reid, it is no surprise that the Coronado does not incorporate the fully developed characteristics of the Shingle Style as McKim, Mead & White, Bruce Price, or John Calvin Stevens practiced it. There is not quite the same sweep of space and surface, but the Coronado is Shingle Style nonetheless, if for no other reason than the two million shingles that cover the enormous roofs. Perhaps half that number again were used in the patterned surfaces of the upper exterior walls.

In portions of the interior, there is certainly an amplitude of scale that suggests eastern shingled work, as in the long, heavily beamed dining room, called the Crown Room, which measures 62 feet down the sides and extends 156 feet (seating 1,000 and a dinner orchestra). The room is paneled and vaulted in sugar pine. The same scale is seen in the sixty-foot-square entry lobby, called the Rotunda, which rises two stories and is paneled and finished in Illinois oak. Originally the floor of the Rotunda was tiled, so that sport fisherman could dump their catch there on returning to the hotel, while their ladies looked on from the safety of the mezzanine above. As the intricate detail of the Rotunda shows, by the time the Chinese carpenters reached this portion of the hotel, their skills had advanced considerably. The various public rooms for card playing, billiards, bowling alleys, Turkish baths, and other facilities, as well as the individual guestrooms, were all appointed comparably to the major public rooms. This was a hotel *de luxe*. Nonetheless, for the 400 bedrooms, there were originally only 73 bathrooms.

Numerous outdoor activities were provided for—there were sailboats, swimming beaches, tennis and golf (then just being introduced to the United States), coach driving, archery, fishing, hunting, and sightseeing trips, with concerts and the Japanese Tea Garden for the more sedate. And all around the encircling balconies were rocking chairs.

Hardly had the hotel been opened in 1888 when the land boom began to falter. By the end of that year the population of San Diego had dropped to 16,000, lots were being sold off as quickly as possible, and the hotel was in financial difficulty. It was saved by an infusion of money from John D. Spreckels, whose family had made a fortune in sugar. Before long Spreckels owned the hotel. Then, in 1899–1900, to bolster the financially troubled hotel, the Coronado Beach Company and the Santa Fe Railroad created a tent city on the sand south of the hotel for travelers with more limited funds. The tent city proved popular among local residents, especially San Diegans, and continued in operation until the outbreak of World War II.

Today the Hotel del Coronado remains a hotel *de luxe*, the grande dame of the southern California coast, where one can still find some of the slow, overstuffed luxury of a Gilded Age resort.

OPPOSITE: *The lobby was grandiosely named "the Rotunda." It retains its original caged Otis elevator.*

FRANK LLOYD WRIGHT HOME AND STUDIO

Oak Park, Illinois, 1889–1914

The living room, inglenook, and hallway are broadly connected yet individuated spaces.

OPPOSITE: *Perhaps the ultimate expression of the dominant front gable first seen in Richardson's Watts Sherman house.*

Vincent Scully's now-classic study, *The Shingle Style: Architectural Theory and Design from Richardson to the Origins of Wright*, concludes with a discussion of Frank Lloyd Wright. It gives Wright's house in Oak Park a place of honor, marking the end of the inventive freedom of the 1870s and 1880s and at the same time announcing the beginning of what would become Wright's Prairie Houses in the early twentieth century.

Wright says nothing in his *Autobiography* about any consideration of Japanese art or architecture in the office of his first employer, Joseph Lyman Silsbee, which Wright entered during 1887. Silsbee, however, was the close boyhood friend and later brother-in-law of Ernest Fennelosa, who was then becoming the foremost American authority on Japanese art and culture. Regardless of the origins of the Japanese influence, clearly Wright was inspired, for in his own house he opened up the rooms to one another, like a Japanese house with the sliding screens pushed back, and he employed a continuous upper molding, running around each room, like the Japanese *kamoi* rail, linking the rooms together.

The most obvious influence on Wright was the East Coast Shingle Style, then being introduced in Chicago by Silsbee, a recent transplant from Syracuse and Buffalo, New York. Silsbee's houses of this period were largely Shingle Style designs, similar to those of eastern architects John Calvin Stevens, McKim, Mead & White, and Lamb & Rich. Silsbee came to the attention of developer J. L. Cochran, who was about to lay out a model suburban community to be called Edgewood, about six miles north of the heart of Chicago. In 1887 he engaged Silsbee to design the houses for this community. Wright, just months in Silsbee's employ, executed a perspective drawing of Cochran's own house from Silsbee's design. Like Bruce Price's houses for Pierre Lorillard in the New York suburb Tuxedo Park, the Edgewood houses were to be relatively small and compact. As in the case of Price, Silsbee was inspired to devise simple dramatic forms in which large dramatic triangular gables predominated.

Wright was aware, too, of the boldly triangular shingled houses being built in Austin, a new suburb just west of Chicago and immediately east of Oak Park, where he lived. Rare photographs survive of the earliest buildings

Wright achieved a unique synthesis of the classical and oriental influences that pervaded Shingle Style design.

there—boldly massed broad-gabled shingled designs by Frederick Schock (fig. 26). A brief mention of Schock in Wright's *Autobiography* suggests that Wright knew these buildings as well. But the most obvious models for Wright's house in Oak Park were Price's shingled houses at Tuxedo Park (fig. 4). Their simple design program encouraged bold, simple, dramatic forms composed of large triangular gables with long sweeping roof lines. One of these houses in particular seems to have been the inspiration for Wright's design: the Chandler house. Its dramatic gable appeared as a linear photoengraving, together with a plan, in *Building* (September 1886).

The changes that Wright made in moving beyond his apparent models anticipate the direction his work would take in the next two decades. As Neil Levine notes in writing about Wright's dramatically abstract Oak Park house, it is the "projection of an image" of what a house could be, at once familiar and yet strikingly simple, and outside the limits proscribed by conventional types. Indeed, Wright comments in the *Autobiography* that his neighbors were perplexed and asked if the design "were Seaside or Colonial."

Wright's first significant innovation was placing his house not on a light framed porch but on a solid elevated terrace, enclosed by a continuous masonry wall and gained by broad low stone stairs, making a far stronger connection to the earth. Wright used continuous surfaces of shingles throughout, on both the walls and long roof planes. He also enlarged and abstracted Price's near-Palladian window, making it a broad strip of windows illuminating his studio. The great overhang of the front gable portends the extended cantilevers of the eaves of Wright's subsequent Prairie Houses.

Wright's plan was a pinwheel of spaces arranged around a small central hearth sheltered within a diminutive inglenook. The round-arched fireplace, with its long tapered brick voussoirs, speaks of Wright's admiration for Richardson and Louis Sullivan. In the four corners of the living room ceiling, electric lighting fixtures are integrated into square-paneled flourishes of foliate ornament, recalling the similarly integrated ornament and lighting used by Sullivan in his Auditorium theater. The staircase in the adjoining entry stair-hall, incorporating a built-in seat and rising in gentle stages with many landings, exemplifies the Queen Anne house. And in the stair-hall, placed over the upper molding, is a continuous plaster frieze, a miniature near-replica of the imposing high relief sculpture of the great Altar of Zeus of Pergamon, whose classical reference is reinforced by the denticulated cornice in the living room.

What began as a compact cottage house was modified repeatedly by Wright to accommodate his family, and then to house his office and studio, so that its original simplicity has been somewhat obscured. Nonetheless, the dramatic west facade gable and the interconnected extruded spaces within still herald Wright's incipient early modernism.

Fairmont Cemetery Chapel

Spokane, Washington, 1890

A tiny but monumental structure of thoroughly unprovincial sophistication.

OPPOSITE: *The linear energies of simply crafted wrought-iron enliven the entry gates.*

"Go west, young man," advised Horace Greeley, and in the last quarter of the nineteenth century, many skilled architects, sons of prominent Eastern families, did just that. Seized perhaps with a spirit of adventure but also seeing professional opportunity, they left comfortable if conventional surroundings to seek their fortunes in the burgeoning cities of the West. The writers of criticism and history, however, remained ensconced in the East, so that much remarkable architecture in the West remained unexamined in the professional press, and hence largely unknown. One of the most colorful of the architectural "missionaries" to the west was Kirtland Cutter (1860–1929), who settled in Spokane, the major city of eastern Washington. Born near Cleveland, Ohio, Cutter came from a prominent family of physicians and businessmen who had roots in Connecticut.

The young Cutter decided to become an artist or illustrator and was sent to study at the Art Students League in New York. Founded by disgruntled students of the staid National Academy of Design, the League encouraged self-reliance and independent thinking, lessons Cutter took much to heart. In the early 1880s Cutter decided to tour Europe and continue his training. Much of his time was spent in Switzerland, Dresden, and Nuremburg, where he was entranced by the romantic but coherent picturesqueness of the old cities. By 1886 he was back in the United States, but his ambitions had been redirected. After learning that he had a color vision impairment, he gave up thoughts of being an artist and decided to become an architect. He had only minimal professional training. Glowing accounts of abundant opportunities in Spokane enticed him westward, where he designed a house for an uncle who had founded the Bank of Spokane Falls.

Almost immediately, Cutter acquired some property of his own and began designing a house for himself, heavily influenced by his memories of Swiss and southern German chalets, and also influenced by Downing's chalet designs. The Chalet Hohenstein, as Cutter's cottage came to be known, was built in 1887 and enlarged several times later. In both his uncle's house and his own, Cutter established a free-spirited bohemian and exotic character, part chalet and part Queen Anne, that appealed to progressive business leaders. It provided a welcome counterpart to the heavy Second Empire Baroque style

Inside the chapel, a starkly beautiful composition of elemental forms.

that was championed by the older Herman Preusse, then the city's most established architect.

Forthwith, two of Horace Cutter's business partners then commissioned Kirtland Cutter to design houses for them in 1889. For the F. Rockwood Moore residence (1889–90), Cutter devised a rambling house, with a rough basalt masonry base and shingled upper stories, accented with half-timbered gables, that Cutter's biographer, Henry Matthews, compares to McKim, Mead & White's Thomas Dunn house in Newport (1877). For the Glover house (1889), Cutter drew upon the broad, multistory gable motif used by McKim, Mead & White and John Calvin Stevens. The base and lower walls were of rough granite, with half-timbering in the wide gable. The entry was recessed in a yawning semicircular arch in the granite base, clearly inspired by Richardson. The interiors of Cutter's houses increasingly exhibited Arts and Crafts qualities and were resplendent with art glass. Moreover, all of these residences, including those for his uncle and himself, were nestled into the sides of hills and integrated into their settings in a novel way. Cutter's loving embrace of the natural setting was unusual and much appreciated by the progressive elite of Spokane. Visitors from the East praised Cutter's designs and soon the young man was an architect to be reckoned with.

During 1889 and 1890 there quickly followed commissions for major business blocks in downtown Spokane, whose rapid design evolution show what can only be called a stunning maturation in Cutter's work. Within months, so it seemed, Cutter had learned all he could from contemporary New York and Chicago examples, from Richardson and George B. Post. The end result can be seen in the diminutive chapel Cutter designed in 1890 for the Fairmont Cemetery.

The Fairmont Chapel has a compact cruciform plan, with a broad roof. The entire design is governed by a restriction in materials and simplicity in design that create a restful, soothing ambience. Externally the materials are extremely rough basalt masonry, with massive buttresses at the corners, and (originally) cedar shingles, graced by the delicate floral tendrils of the wrought-iron entry gate. The entry arch, like that of Richardson's Ames Gate Lodge in North Easton, Massachusetts, gently pushes up against the shingles above, its arc generating concentric ripples in the shingles. The angular sharpness of the rough basalt walls is countered by the whimsical curves of the roof that rise to playful cupolas. The dark interior, its severe lines softened by the curvilinear patterns of wrought-iron gates, is dominated by the glow of the circular art glass window, reiterating the cross pattern, with its words of solace: "I am the resurrection and the life."

ERNEST COXHEAD'S HOUSE

San Francisco, California, 1893

The fireplace at the rear of the long gallery.

OPPOSITE: *Winding flights of steps lead to the front door.*

Architecture "on the edge of the world" was what architectural historian Richard Longstreth called the work of several highly imaginative architects who moved to San Francisco at the turn of the last century. Almost at once that city was blessed with the inventive genius of five remarkable designers—Ernest Coxhead, Willis Polk, Bernard Maybeck, A. C. Schweinfurth, and A. Page Brown. All came from the East. Maybeck had worked in New York City in the office of Carrère & Hastings; and Brown for McKim, Mead & White.

Ernest Coxhead, however, came from much farther east. Born in 1863 in Eastbourne, Sussex, England, Coxhead had studied under an engineer and then at the Royal Academy and the Architectural Association in London. Thanks to his work and education Coxhead possessed a solid grounding in classical design, with its emphasis on clear expression of the building program and its emphasis on proportions, as well as a sound introduction to English medieval architecture, with its attention to detail. He was involved in the restoration of several centuries-old churches and seems to have developed some associations with the young leaders of the English Arts and Crafts movement in London. In 1886 he and his brother, Almeric, left Great Britain and headed west, crossing the American continent and settling first in Los Angeles, California. Why he made so decisive and dramatic a break from family and country may never be known, but he may have been given encouragement by the Episcopal Diocese in California. Between 1887 and 1898 he and Almeric, who managed their practice, designed most of southern California's new Episcopal churches and enjoyed a field of action far greater than would have been afforded them in England.

While in England Coxhead had been introduced to the American Shingle Style. Longstreth notes that a major exhibition of such American work was mounted by the Royal Institute of British Architects shortly before Coxhead left. One of Coxhead's early churches, All Saints in Pasadena, 1888–89, employed a fusion of English Arts and Crafts with the rounded, biomorphic forms made possible by shingle work. Other churches followed, but the building boom in Los Angeles ended in about 1889 as Coxhead was given commissions for three new Episcopal churches in the San Fransicso Bay area.

ABOVE: *Eschewing symmetry and formality, Coxhead made his living room a collage of cozy corners.*

His first project in San Francisco, and perhaps his masterwork in church design, was the massive Church of St. John the Evangelist, 1890–91 (fig. 28). It was dynamited to prevent the spread of fire following the earthquake of 1906. Indebted to Richardson, it was based on a compact Greek cross plan but had a center dome capped by a broad squat square shingle-covered tower, vented by deep louvers that ran in continuous bands around the base of the pyramidal roof. The shingled roof surface also wrapped over the gable ends, fusing with the wall surfaces in a unique organic way. Although his other major urban churches were of masonry, Coxhead's smaller parish churches exploited shingles, which seemed to flow over the building surface, around corners, up and over doors and windows, and over gable ends, merging wall and roof into one plastic envelope.

By 1891 the Coxhead partnership began to receive commissions for small houses in San Francisco, such as that for James McGauley on Pacific Heights. For these Coxhead continued to use wood frame construction, and in the McGauley house he used an exposed half-timber frame, interrupted by a broad brick chimney mass, and a tall, steep roof that prompted Longstreth to call the house a "transplanted English cottage." By 1893 Coxhead's house designs had become more abstracted, their geometric shapes emphasized by continuous coverings of shingles over the walls and roofs. Windows were grouped and placed strongly off-center at what appear to be odd locations but which actually reflect the pragmatic arrangements of the interiors. In some instances, the unusual character of these houses was dramatized by curiously overscaled details. Certainly, a contributing factor in Coxhead's distinctive work were the steeply pitched building sites he worked on, as in Pacific Heights, for the front facades of the houses would automatically be thrown off center by the incline of the street.

In 1891–92, adjacent to the McGauley house, Coxhead designed an extremely long and narrow house for himself and his brother. The narrow street facade, rising four stories, becomes almost a tower, while the entry side (reached by steps and a tunnel-like passage through the base retaining wall), stretches almost 94 feet, with the steep roof plane pulled deliberately low to

At the rear of the long gallery.

ABOVE: *With the door closed, this corner of the bedroom becomes an intimate sitting area.*

OPPOSITE: *The tiny staircase demonstrates Coxhead's skill in turning the exigencies of a narrow lot to picturesque advantage.*

emphasize its horizontal extension. The narrow site gave rise to some unusual innovations, such as a long entrance corridor that Coxhead broadened a bit to evoke memories of an English long gallery. With two hearths introduced, this gallery divides itself into separate sitting areas. The rear area is especially pleasant. A bay window and French doors bring in abundant light even on gray, foggy days. At every turn the exigencies of the narrow site, and the low roof, are turned to advantage to produce unexpected nooks and cozy recesses. Dark wood, broadly and blockily detailed, dominates the interior spaces, further bringing down the scale. Although dark and encompassing, the rooms are opened up by broad window groupings, which once afforded panoramic views of San Francisco Bay. As neighboring buildings began to impinge on his views, Coxhead moved away, but his rustic aerie survives, an enchanted little world of domestic delight.

ST. JOHN'S CHURCH

Petaluma, California, 1890–91

Ernest Coxhead was a well-established designer of churches when he and his brother Almeric moved from Los Angeles to San Francisco around 1890. He had already completed eleven churches, including the Church of the Angels just outside Los Angeles, in 1889, and All Saints Episcopal Church in Pasadena, in 1888–89. In his Chapel of St. John the Evangelist, in Monterey (1891), he began to use shingles in highly innovative ways, rounding corners and intersections, softening the outline of the building. All these churches emphasized traditional liturgical organization, with a clearly distinguished chancel and nave.

Coxhead's few articles on church design reveal a dedication to the High Church Anglican ideals he had absorbed in England. He largely disapproved of the American tendency to build churches like auditoriums to emphasize the sermon. Ironically, his largest completed church, St. John the Evangelist in San Francisco (1890–91), was just such a centralized auditorium church, sheathed externally in undulating and sinuous shingle surfaces. Over the

RIGHT: *In a typically Coxhead sleight of scales, two tiny vestibule windows make the front door seem even more massive.*

OPPOSITE: *A blanket of shingles envelops the church.*

In a modest country church, Coxhead created an impression of vast, cavernous space.

years, Coxhead designed approximately twenty-five churches, of which eighteen survive. Many have been modified or otherwise compromised, but his St. John's Episcopal Church, Petaluma, survives nearly intact.

Petaluma in the 1890s was a prosperous small town in Sonoma County, thirty miles north of San Francisco. Its economy was primarily agricultural, supplying crops and poultry for San Francisco. At the edge of the coastal mountain forests, it also served as a lumber shipping port. In 1890 the Parish of St. John's needed a larger church, and plans were prepared by Coxhead in San Francisco.

Although St. John's has a liturgical plan, its orientation is not orthodox and does not run west to east. The axis of the plan runs on an angle north to south, with the smaller choir and chancel areas to the south, at the back of the site. Entry to the church, through massive oak doors on large wrought-iron decorative hinges, is through the base of the prominent tower at the intersection of Fifth and C Streets. The curved north wall, seeming to be an apse, in fact provides a recessed space housing the baptismal font at the back of the nave and adjacent to the entry. Since baptism symbolizes entry into the church, the font is near the principal doors, as in the English tradition. The liturgical focus on the altar is emphasized by the low side aisles along the nave; the separate choir space raised three steps (with stalls running parallel to the central aisle, as in English Gothic churches and chapels); and a recessed, narrower chancel with altar, set off by a lowered barrel-vaulted ceiling. The nave itself has a steeply pitched gable ceiling. Small clerestory windows of stained glass, and stained-glass ogee-arched windows in the side aisles, admit a low, mysterious light.

Outside, Coxhead's playful use of shingles is wonderfully evident. Except for the stone at the foundations and in the base of the corner tower, the church is entirely swathed in a blanket of shingles. They swell out where the wall meets the stone base, and they bend around corners, curving around the planes of the spire and corner tourelles, around the ventilating dormer, and over the curve of the north wall. Perhaps no other architect so emphasized the continuous character of the shingled surface. At important junctures, such as at the bottom of the shingled walls, or where the eave of the roof meets the tower, or where the deep louvers of the belfry intersect the upper walls, there the rows of shingles are cut in sharp diamond teeth, which make delicate continuous lines that wrap the building. Where there are large unbroken expanses of wall, a large diamond pattern is introduced.

Over the arched opening in the tower base is a semicircular pediment, almost baroque in its cut-away base and crowning volutes, filled with low relief carving and a quatrefoil window that recalls such Mission churches as San Carlos Borromeo in Carmel. All of this rests on a very unclassical profusion of Ionic columns. This flourish of applied decoration on an otherwise restrained building is characteristic of the quirky genius of Ernest Coxhead.

FELSTED

Deer Isle, Maine, 1896

Gently angled wings and a broad front door welcome visitors to the Olmsted family cottage.

OPPOSITE: *Connecting porches meander around the water side of Felsted.*

OVERLEAF: *The cottage rises from a massive granite base to command a rugged stretch of the Maine coast.*

It is easy today for casual visitors to Prospect Park in Brooklyn, or the town of Riverside, Illinois, to be underwhelmed, to think that the landscape they see is simply a remnant of nature as it once was. The plantings seem so natural and fitted to their place that one is surprised to learn that virtually every tree and shrub was carefully selected by landscape architect Frederick Law Olmsted to create an environment shaped to nourish and restore human sensibilities. One can hardly imagine what the land looked like before Olmsted set to work.

Much the same can be said of the house initiated by Olmsted's wife, Mary, as a place of restorative calm for her husband. Perched atop a granite outcrop, overlooking the rippling expanse of Maine's Penobscot Bay, it seems to have always been there. One can hardly imagine the rock without the house, so fused are they into an integral whole.

The house was commissioned by the Olmsteds from architect William Ralph Emerson in 1896, when the Shingle Style was waning along the Atlantic seaboard. Yet in this house Emerson seemed to sum up and give special force to the type of summer house he had been instrumental in inventing in the late 1870s. Olmsted's health was seriously failing—he was now seventy-four years of age, spent by years of travel and work in the field—and this house was intended to provide a quiet place where the sea air might help to reinvigorate the aging landscape designer. Olmsted's son, Frederick, Jr., personally directed the building campaign, writing to Emerson in April 1896, to arrange a site visit. In late April the two visited Deer Isle, off the beaten path but not too distant from Bar Harbor, where the Olmsteds had stayed during previous summers. They agreed in principle on a "low and rather straggling house," generally small and compact, with a bedroom on the main floor for Frederick's father, an adjoining smaller bedroom for an attendant, another chamber for Mary Olmsted, a bedroom for a doctor or guest, and the usual living and dining rooms, plus a kitchen and pantry. The name for the house was an abbreviated version of Olmsted's own.

The final plan for Felsted was shaped by Emerson in the form of a stubby Y, with the living room (and a bed chamber above) thrust out over the water, and the living room partly surrounded by a covered balcony that ran around the house to the kitchen. The bend of the Y allowed for sweeping views from

The view of Penobscot Bay from the living room porch.

each major room—Olmsted's bedroom, the living room, and the dining room—largely unobstructed by other parts of the house. At the end of Olmsted's bed chamber is a separate covered balcony. Altogether there are thirteen rooms, with fireplaces. Yet it is not a large house, and all the rooms except the living room are small. Because of the bend in the plan, the rooms bump up against each other, creating intriguing angles and intersections. Simply plastered with a wainscot of greenheart, a South American hardwood, the house has an unpretentious and comfortable character. Compared to other Shingle Style cottages—enormous mansions in informal summer attire—Felsted is cozily intimate, but with a view whose reach to infinity provides the spaciousness.

Felsted is lifted up on a stone basement, with engaged piers carrying massive brackets that support the encircling balcony and porch. The porch posts are fitted with thick kneebraces, giving the house a sense of permanence. The subtle curves of the brackets and kneebraces are echoed by the outward curving extensions of the wall carrying the sheltering gable over the upper bedroom. The house is covered entirely with shingles, curving out from the upper wall surfaces to the delicate white cornice that runs around the entire eastern sea side, forming here a lintel over the window, there the edge of the porch roof.

Once a small hotel in the 1920s, a private summer house since 1941, and used as the principal location for the Mel Gibson film *The Man Without a Face* (1986), this secluded house, still privately owned and lovingly maintained, now receives a trickle of admiring visitors—students and scholars of Emerson, Olmsted, and Shingle Style houses—drawn by its magic appeal.

OPPOSITE: *The major rooms all face the water, and are flooded with abundant light and spectacular views.*

FIRST UNITARIAN
CHURCH

Berkeley, California, 1898

A.C. Schweinfurth is among the most overlooked of the young architects who brought the Shingle Style to the San Francisco Bay area. The Unitarian Church he designed for Berkeley shows his creative prowess. If not for the ravages of time and the neglect of its present owner, the University of California, this structure could easily be mistaken for a supremely accomplished work of contemporary Postmodern architecture.

Albert Cicero Schweinfurth (1864–1900) was one of four brothers, all of whom became architects. His older brothers Charles and Julius would become well-known architects in Cleveland and Boston, respectively. The interest of the brothers in architecture was certainly encouraged by their father, Charles J. Schweinfurth, trained as an engineer in his native Germany but forced to emigrate after the abortive revolution of 1848–49. The father settled in Auburn, New York, setting up what became a large woodcarving shop that specialized in architectural ornament. After studying with his father, Albert worked in the office of Peabody and Stearns in Boston, where

RIGHT: *A primordial composition of simple shapes and rough materials.*

OPPOSITE: *The few touches of ornament were derived from the materials themselves.*

140

Both ends of the broad frontal triangle were hollowed out to serve as covered porches at the doors to the church.

his brother Charles was also employed. In 1884 Albert moved to New York to work for A. Page Brown, a recent graduate from the office of McKim, Mead & White who, in mid-1889, moved west to set up a small branch office in San Francisco. The California branch office, however, soon became Brown's principal office, and in 1895 Schweinfurth joined Brown there. Schweinfurth's practice lasted only three years; in 1898 he departed for two years of study and travel in Europe. Shortly after his return, Schweinfurth died of typhus. Hence the startling Unitarian church was his last work.

From both Robert Peabody (and McKim, Mead & White, by extension, through Brown), Schweinfurth developed an appreciation of the vernacular shingled buildings of colonial New England. He also was affected by Peabody's response to the enormous wave of Richardsonian influence sweeping the continent in the late 1880s. Schweinfurth's Romanesque design for a Museum of Historic Art at Princeton University, 1888, came directly from Peabody's design for the Museum of Fine Arts in St. Louis. Schweinfurth's scheme is much simpler, however, dominated by an enormous low-slung semicircular entry arch in the center of the building. Schweinfurth must also have known of Brown's design for the large shingled Crocker Old People's Home in San Francisco, 1889–90.

In his independent work, begun about 1895, Schweinfurth gave clear evidence of developing a new regional California architecture incorporating elements derived from Spanish vernacular adobe architecture and even the pueblo housing blocks of the Southwest. He was successful in winning from Phoebe Hearst the commission for a large family retreat, the Hacienda del Poso de Verona, near Sunol, California. His office block for the San Francisco *Examiner*, built in 1897–98 (and destroyed by fire following the earthquake in 1906) was an urbane variant derived from the same sources. But Schweinfurth did not abandon the Shingle Style; his rustic shingled house for James Bradford, San Francisco, 1896, was an extraordinarily bold composition, comprised of a large gable set perpendicular to the street, and broken only by two simple rectangular windows. This front gable, sheltering a recessed porch beneath it, was carried on four huge tree trunk porch posts carrying equally stout square beams whose projecting ends formed dramatic accents under the lower gable edge. The rough surfaces of the bark-covered porch posts were continued in the unbroken covering of shingles on the roof and walls of the house.

This commanding composition for the Bradford house was Schweinfurth's origin for the equally remarkable Unitarian Church. This, too, is a broad gable, lower and more all-encompassing, rather like the enormous low gable of McKim, Mead & White's Low house in Rhode Island, 1886–87. As in the Low house, the ends of the sweeping gable cover recessed porches, but Schweinfurth's massive crossed and interlocked square beams are supported by stout bark-clad tree trunks. The punctuation of the projecting beam ends is picked up in the smaller protruding ends of purlins that project from under the edge of the roof. Running over everything, roof and walls, is a blanket of shingles. A broad band of sawtooth shingles, laid in courses six layers deep runs over the porch beams, turns upward, and runs over the curve of the single large round sanctuary window.

The bark-covered columns, the clear expression of the support beams, and the emphatic triangular gable seem to return this small building to the very origins of architecture. Such a building—as Vitruvius wrote in the first century B.C., and as Marc Antoine Laugier would later reassert in the mid-eighteenth century—is the essence of architecture, primal and pure: post, beam, and roof.

BELOW: *Large windows filled the church with light.*

GIGNOUX COTTAGE

Portland, Maine, 1905–06

The Portland, Maine, architect John Calvin Stevens is best known for his important trend-setting shingle houses of the 1880s, but with his partner, Albert Cobb, he continued to use the shingled idiom for many houses later in his long career. These are beginning to become better known, especially since the publication of Earle Shettleworth's recent book, *John Calvin Stevens: Domestic Architecture, 1890–1930*.

Many of Cobb and Stevens' best shingled houses were built in a summer colony called Delano Park, begun in 1885 on the Cape Elizabeth shore, just south of Portland. Of the initial twenty-five houses, four were designed by Stevens. The firm's Charles Brown house, illustrated by Cobb's drawing in the introduction (fig. 19), is among these but has been altered. Another fourteen lots were added to the park in 1901, and four more houses were designed by Stevens. The Gignoux cottage was the largest in this second group.

Frederick E. Gignoux was a career army officer who had resided in New York but who resettled in Portland after his discharge in 1902. Constructed by builder William Murray, the Gignoux cottage sits atop the highest rise in Delano Park, giving it ocean views on three sides, although these are now diminished by the unchecked growth of mature pine trees. The main block of the house is a rectangle capped by a double-cross-gable roof with greatly

RIGHT: *The carriage house echoes the sloping roofs of the main cottage.*

OPPOSITE: *Double gables were a favored motif in John Calvin Stevens' shingled houses.*

The entrance hall doubled as an informal living room, with a quieter library secluded beyond the sliding doors.

OPPOSITE: *The panelling of the main rooms could not be much simpler, or more elegant.*

OVERLEAF: *On the front porch.*

extended eaves, and a broad pent-roof on the long side covering a wide encircling porch supported by stout pairs of square posts with curved kneebraces. The effect is to make the porch seem quite sturdy, a spacious place to sit in wicker chairs and enjoy the sea breezes. Projecting from the main sections of the house is an angled service wing.

The interiors, with the principal rooms wainscoted in long vertical panels, have a high-waisted character, with a molding running continuously around each room just atop the door jamb, forming a plate rail in the dining room. Here, perhaps in response to the lower, cooler light level of Maine's misty, fog-shrouded days, the interiors are painted white and pale cream, making for more expansive, brighter rooms.

The carefully composed asymmetries of the main house are reiterated in the carriage house, with its equally assured groupings of gables and windows. The long sloping roof to the left of the heavily framed carriage door evokes the profiles of New England's venerable saltbox houses.

GRAYOAKS

Ross, California, 1906

The upper terrace, outside Maybeck's ballroom addition.

OPPOSITE: *Maybeck was long enamored of the simple shapes and great roofs of the Swiss chalet.*

Of the eastern-trained architects who developed the Bay Area Shingle Style, perhaps the most audacious was Bernard Maybeck. His originality and his whimsical distortions of architectural norms made him an inspiration to later generations of Bay Area architects, such as Joseph Esherick. Maybeck's buildings, whether private residences, churches, or educational buildings, combined a vernacular shingle tradition, selected classical or medieval details, and industrial materials and building techniques. Classically trained, Maybeck was an architect who sought to create dramatic and expressive "beauty" but he was also a craftsman-builder, keenly interested in exposing the process of construction. Maybeck also had the good fortune to live to a ripe old age, so that young architecture students at Berkeley, such as Kenneth Cardwell, were able to learn first-hand in later years what Maybeck had attempted to do in his work.

Bernard R. Maybeck, born in New York City, was trained as a cabinet-maker by his father, a recent immigrant from Germany. The elder Maybeck encouraged his son's avid freehand drawing, and soon the young man was sent to Paris to work in the furniture shop of Pottier and Stymus, the parent company of his father's employer. Working very near the Ecole des Beaux Arts, young Maybeck saw the architectural drawings students brought to the school and decided that he, too, wanted to become an architect. He was admitted to the Ecole, and after five years of study he left in 1886, returning to New York. There he entered the office of Carrère & Hastings, who themselves had only recently been assistants in the office of McKim, Mead & White. Hence, Maybeck not only had the benefit of the most theoretical architectural education in Paris, with its emphasis on formal design, but he was also exposed in New York to the fully developed eastern Shingle Style, then reaching its creative zenith. Maybeck produced a few shingled residential designs at Carrère & Hastings before he decided to seek an independent career in Kansas City. Hearing of even more promising professional opportunities in California, Maybeck and his bride resettled in San Francisco in 1890.

While developing his architectural practice in San Francisco, Maybeck was engaged in 1894 to teach drawing at the University of California in Berkeley. He soon became acquainted with the great patroness of the university, Phoebe Hearst. For the Hearsts, Maybeck designed a gymnasium for the

Simple redwood panelling glows in the light of the living room.

university and a romantic family retreat called Wyntoon in the Sierra Nevada mountains. He also began to receive commissions for modest frame houses in the hills above Berkeley and Oakland. These houses exploited their tight, irregular lots with innovative plan arrangements. They were capped with steep roofs and their interiors were marked by exposed wood framing and simple paneling. In most of them, the exterior wall and roof surfaces were covered with redwood shingles.

Maybeck was especially interested in dramatic visual effects and striking compositions, using bold roof overhangs and incorporating oversized Gothic details, as he did later in his much admired First Church of Christ Scientist in Berkeley, 1910. He also took a very serious view of residential design and spent long periods of time with his clients, learning their needs and desires. In an interview in 1927 Maybeck said, "the thing to do is make the home fit the family. . . . I never plan a home for a man until I have asked him a lot of questions. 'What sort of woman is your wife? What kind of clothes do you both wear? What do you most like to read? Do you enjoy music?'"

In sites remote from crowded Berkeley, Maybeck was even freer to develop unusual new plan arrangements and experiment with building materials. His country houses took on something of the character of chalets, but significantly, when building on a slope Maybeck always preferred to place the house with its long dimension parallel to the hill, opening up the longer facade to views. This was how he positioned "Grayoaks," the country house of J. H. Hopps, a lumber baron who owned a large expanse of rolling, oak-studded land in Marin County, across the Golden Gate from San Francisco. Now screened by tall trees, the house once enjoyed a sweeping view across the Ross Valley to Mount Tamalpais.

A relatively simple box with an enormous overhanging roof, with eaves supported by huge brackets, the house was sheathed in oversized redwood shingles, three feet long, laid with a short distance between them to create a striking pattern. As elsewhere, Maybeck played with contrasting materials, notably in the chimney with its rough stone lower section supporting an upper stage of brick. As in his other earlier houses, the cabinetry is shorn of extraneous details and relies on the grain of the clear old-growth wood to provide visual texture. In 1925 Maybeck was brought back to expand the house with a perpendicular wing to the rear, a ballroom for the wedding of one of the daughters, creating a semi-enclosed court. The addition, however, is clad in stucco, setting it apart in time from the original section. The double door of the new section, with its large plate glass transom, looks particularly modern. Yet a step inside the door reveals a Tudor arch at the great hearth.

Maybeck's architecture, whether private or public, residential, commercial, or religious, always contains surprises just around the corner, created through unexpected historical allusions, materials used in novel ways, dramatic changes in scale or in light levels. The Hopps house is an early harbinger of these qualities so characteristic of his long career.

In the ballroom addition, Maybeck
combined his enthusiasms for redwood
panelling and for abstracted gothic details.

154

GAMBLE HOUSE

Pasadena, California, 1908–9

Wonderful though it is, the Gamble House may not immediately come to mind as an example of a shingled building, because its interiors and details are so decidedly Arts and Crafts. Moreover, a glance at the exterior may register the deeply shadowed voids of the sleeping porches, punctuated by the assertive articulation of the framing, rather than the smooth expanses of shingled surfaces. Nonetheless, this house, too, is an integral part of the national shingled architecture. Its emphatically horizontal design, its easy flow of interior space, and its broad embrace of the surrounding landscape mark it as a descendant of the eastern Shingle Style.

At the turn of the century, easy railroad access brought many to southern California for health, rest, and relaxation. Eleven miles northeast of Los Angeles, near the base of the San Gabriel Mountains, what would become the town of Pasadena was settled in 1874 by fruit growers from Indianapolis. Many other midwesterners would eventually follow, exchanging their bitter winters for the famously mild and sunny climate of Pasadena.

The wealthy built palatial homes along South Orange Grove Boulevard, popularly known as "Millionaires' Row." In southeast Pasadena, in what became known as Oak Knoll, another cluster of large winter homes was built, among them the spacious shingled house of 1907 for Robert R. Blacker by Greene & Greene. Shingled bungalows became popular all over

RIGHT: *The house fronts a broad lawn and a gated street. Greene & Greene designed the garage to complement the house.*

OPPOSITE: *An art glass window overlooks the extravagantly crafted staircase, allowing callers to be seen from the master bedroom.*

Pasadena, and many were rented to winter visitors. These winsome structures often appeared in illustrated essays on bungalow design in *Ladies' Home Journal* from 1904 to 1910.

The most celebrated bungalow was built by David and Mary Gamble of Cincinnati, Ohio. David Berry Gamble was a second-generation member of the Proctor and Gamble Company and had retired in 1895. For a number of years the family had wintered in Pasadena. Seeking a warm and healthy place to settle, in 1907 the Gambles bought property in northwest Pasadena on Westmoreland Place, on a slightly rounded knoll near the edge of the Arroyo Seco ravine, looking over the valley where the Rose Bowl is today.

The Gambles selected architects Charles and Henry Greene, who had developed a particular style of house that emphasized broad, low spaces, with shade from the warm sun, abundant cross-ventilation, easy access to outdoor terraces and patios, and a close relationship between house and landscape.

Charles Sumner Greene (1868–1957) and his brother, Henry Mather Greene (1870–1954), were born fifteen months apart in Cincinnati but were raised in Wyandott, Virginia, where their father was studying medicine. The family moved to St. Louis, where the father practiced medicine. The boys studied at the Manual Training High School run by Washington University, where they had extensive hands-on training in building and cabinetmaking. In 1888, the young men, who both aspired to careers as architects, were sent east to MIT. Upon finishing their studies in 1890 they worked in Boston architectural offices, notably that of Shepley, Rutan & Coolidge, successors to the practice of Henry Hobson Richardson.

Thus the brothers were in Boston at the zenith of the eastern Shingle Style. What they absorbed from Richardson's influence was a feeling for powerful scale, a concern for clearly expressed structure, and a preference for the direct use of materials. Meanwhile, their parents, both in failing health, had moved to Pasadena, and in 1893 they urged the boys to pay them a visit. En route, Charles and Henry stopped in Chicago to tour the Columbia Exposition and were deeply impressed by the Japanese pavilion, with its carefully detailed timber construction.

The visit to California led to a decision to settle and make their careers in Pasadena. While studying local architectural traditions, the brothers at first produced houses derived from the Eastern models they knew well, such as Queen Anne shingled houses. But they became increasingly influenced by the Arts and Crafts movement, with its attention to the integrated design of houses, interiors, details, and furniture. The framing of their houses became more strongly articulated, with roof planes extending into broad shady eaves. Walls were usually covered with shingles, the framing and shingles set off by contrasting foundations of brick and local arroyo stone. An early example of their mature work is the Darling house of 1903 in Claremont, California. Rather modest and only two rooms deep, it is sheltered by an ennobling broad gable roof, combining aspects of McKim, Mead & White's Low house with Swiss chalet elements. It was also one of the Greenes' first designs to

The dining room table reflects amber light from stained glass windows over a built-in sideboard.

In the living room, the frieze depicts oriental landscapes. Here, an evening sky with bats and a full moon.

receive international attention; it was published in *Academy Architecture* in London in 1903.

The Greenes built a colony of modest shingled houses from 1900 to 1907 along Grand Avenue and Arroyo Terrace in Pasadena. Among these was Charles Greene's own home, begun in 1901 and expanded in sections through 1916 to accommodate his growing family. Immediately north of this cluster of houses, the Gambles acquired their large parcel of land in 1907.

The Greene brothers had plans ready for construction of the Gamble House in February 1908, ground was broken in March, and construction continued until February 1909. The house was placed on a low rise, facing a curving brick driveway. Tapering brick steps lead up to the tiled front terrace. The broad entrance door is the middle section of a screen of art glass, which opens to a broad hall that runs back through the main floor of the house. To the left is the working section of the house, with a guest bedroom and service quarters. To the right is a winding stair, gently rising in a series of landings. Further along the central hall are wide openings to the dining room on the left and the large living room on the right.

Large doors and windows open the back of the house to a terrace of tile and brick. The terrace gradually shifts from a rectilinear grid corresponding to the house, to curving lines of brick which wrap around a curvilinear planting bed where two great eucalyptus trees once stood. The heavily battered retaining walls of the terrace are built of clinker bricks inset with irregular rocks, seemingly more a work of nature than of master masons.

No detail of construction was left to chance or happenstance. One of the brothers was on the job site nearly every day, closely supervising the work of builder and master craftsman Peter Hall. The brothers personally selected rocks from the canyon below for the retaining walls. They designed every element and fixture in the house, such as the stained glass windows and light fixtures, which were fabricated by Emile Lange from pieces made by Tiffany Studios. Lange's greatest work in the house was the entry screen, with its oak tree design unifying all three sections. The Greenes' furniture designs were executed in the shops of Peter Hall under the direction of Peter's brother, John. They included a special mahogany case for the living room's upright piano. The Greenes also devised the patterns for the specially woven rugs.

In addition to the influences of the Arts and Crafts movement, seen in the art glass and custom furniture, the Gamble House was shaped by the Greenes' abiding interest in Japanese art and architecture. Throughout the house, there appears a motif called the "cloud lift," based on Japanese screen painting, in which the line shifts and rises gently. The cloud lift can be seen at large scale in the open beams over the living room inglenook, and at smaller scales throughout the structural details and furniture of the house, even in the cames of the light fixtures. Also drawn from Japanese architecture are the softly rounded ends of every wooden structural member.

ABOVE: *The family bedrooms have expansive sleeping porches. Sleeping out-of-doors was believed to be a therapeutic benefit of the balmy Pasadena climate.*

OPPOSITE: *The terrace walls are built of clinker brick, inset with stones that the Greenes selected from the nearby arroyo.*

OVERLEAF: *Twilight on the terrace.*

The Greene brothers refused to simply turn over drawings to builders or contractors. Their hands-on involvement meant that unsatisfactory work was often ripped out and done over. Ultimately the brothers priced themselves out of their market. By 1916 most of their houses had been built; that year the personal income tax was imposed, and a good deal of discretionary income was redirected. Vacation habits changed, too, as the wealthy sought out tropical environs less staid than Pasadena. Fortunately, several generations of Gambles cherished their house. Without any significant changes, the house remained in the family until 1966, when it was given to the city of Pasadena in a joint agreement with the University of Southern California. Today it is open to the public. It is a supreme monument to a high-minded family and to architects and craftsmen with unwavering commitment to excellence in design and workmanship.

"THE AIRPLANE HOUSE"

Woods Hole, Massachusetts, 1911–12

Light fixtures crown a sideboard built into the entrance hall.

OPPOSITE: *The lower stair landing.*

OVERLEAF: *The bungalow was dubbed "The Airplane House" because it resembled a biplane poised to soar over Woods Hole.*

In 1911 Harold and Josephine Crane Bradley commissioned a seaside bungalow at Woods Hole, Massachusetts, from the Minneapolis architectural firm of Purcell & Elmslie. The Bradleys knew the work of Elmslie, who had supervised the execution of a house in Madison, Wisconsin, that Josephine's father, Richard Crane, had commissioned in 1908 from Louis Sullivan. Lovingly maintained today, the Bradley seaside bungalow, popularly known as "The Airplane House," represents the migration of the Shingle Style to the Midwest, its cross-pollination there with the newer Prairie School, and then the boomerang-like return of this hybrid, updated Shingle Style to its original birthplace, the East Coast.

George Grant Elmslie had been Louis Sullivan's steadfast associate during the gradual decline of Sullivan's practice around the turn of the century. A native of Scotland, Elmslie immigrated at age thirteen to Chicago with his family. He attended schools in Chicago and in 1888 entered the architectural office of J. Lyman Silsbee, where he became acquainted with Frank Lloyd Wright, who soon left for the office of Adler & Sullivan. When Wright left Sullivan's office in 1893, Elmslie became Adler & Sullivan's chief draftsman. Following the dissolution of Sullivan's partnership with Adler in 1895. Elmslie remained with Sullivan until, with the absence of any commissions, he finally had to leave in 1909. Elmslie moved to Minneapolis to join the young William Gray Purcell, whom he knew from Sullivan's office. Purcell, a native of Oak Park, Illinois, who had studied architecture at Cornell University, worked very briefly in the office of Henry Ives Cobb and then briefly for Sullivan in Chicago during 1903.

The dramatic cruciform plan of the Bradley seaside bungalow is derived from the Madison house, where Sullivan and Elmslie had also used a crossed-axial plan. The bungalow is made of two linear elements, the upper element crossing at right angles over the lower element. The ground-floor plan is T-shaped, with service quarters and kitchen in the tail of the T, its head consisting of the great semicircular bay. Inside the great bay is a broad living room focused on a central fireplace mass placed at the crossing of the two axes of the bungalow. The crossbar of the T-shaped ground-floor plan is comprised of two covered porches, which are sheltered by the bedrooms overhead and their sleeping porch extensions.

Once, the upper story of the bungalow truly appeared to float, for its sleeping porches were enclosed only by screens, and the lower porches were enclosed not at all, allowing one's gaze to pass through much of the house. The bedroom section's resemblance to the wings of a biplane occasioned the nickname by which the bungalow has been known ever since. The eventual glazing of the porches compromised this effect, but the cantilevered sleeping porches and their greatly extended eaves still impart a sense of soaring lift.

Emphasizing this sense of flight is the placement of the bungalow, alone, on the crown of a hill near the end of a narrow spit of land jutting out into the harbor of Woods Hole; it stands splendidly isolated against the sky. In planning the bungalow, Purcell & Elmslie emphasized this incomparable setting, keeping the side porches free of interruptions except for the piers supporting the bedroom wing above. They designed ten pairs of art glass windows around the arc of the living room, to give a modulated view of the ocean horizon beyond. Facing the windows is the broad mass of the fireplace, built of salt-glazed, tawny-colored Roman brick. Its yawning semicircular opening echoes in elevation the great curve of the room in plan; its long voussoirs reveal the long-lived influences of Richardson and Sullivan. The warmth of the fireplace brick is matched by the woodwork and cabinetry of the interior, and by the amber-colored glass of the lighting fixtures.

The long dimensions of the flat Roman brick are characteristic of the pervasive horizontality of the bungalow. On the upper floor, the snugly functional bedrooms resemble staterooms on a ocean liner, reached by a narrow corridor of great length. Even the shingles of the exterior walls stress the horizontal line, with every fourth row laid with a very short exposure to the weather. This emphasis of the horizontal line had emerged in early eastern Shingle Style works such as the Newport Casino, and thereafter became a fundamental design strategy of Wright and his Prairie School colleagues. With the Bradley bungalow at Woods Hole, the horizontal expansiveness of the Shingle Style found dramatic, updated expression on the shores of the Atlantic.

The brick mass of the fireplace anchors the "Observatory," Elmslie's name for the living room.

John Galen Howard House

Berkeley, California, 1912

Among the adventurous young eastern architects who settled around San Francisco Bay at the turn of the century was John Galen Howard. Howard had learned the Shingle Style at the source. Born near Boston in 1864, he studied architecture at MIT but left before graduating to work in the office of Henry Hobson Richardson. Seeking work in the office of Ernest Coxhead, Howard journeyed west to Los Angeles in 1888. Although he enjoyed the California climate, he disliked the boom-town disorder of Los Angeles and returned east to undertake a grand tour of Europe. He returned to New York in 1889 and worked for McKim, Mead & White. At McKim's urging, Howard set out for the Ecole des Beaux Arts in 1891; he studied in Paris for two years and then settled in New York, forming a partnership with S. M. Cauldwell. His design for the Electric Tower at the Pan-American Exposition in Buffalo, New York, 1901, was much admired.

At this point Howard's career took a decisive turn. Like many young architects, then and now, Howard entered design competitions in the hope

RIGHT: *The house turns the street corner with aplomb.*

OPPOSITE: *A light-filled stair hall links all the rooms on the main floor.*

A hushed patio is nestled between the house and its hillside garden.

of dramatically advancing his career. In 1899 he submitted a design in the competition sponsored by Phoebe Hearst for a campus plan for the University of California at Berkeley. His submission was awarded fourth place; the competition was won by the French architect Emile Bénard, who proposed a sweeping axial composition dotted with imposing classical buildings. But when it came time to implement his plan Bénard refused to leave Paris for what he must have taken to be the absolute fringe of civilization. So the regents of the University offered Howard the dual positions of supervising architect and professor of architecture. So it was that Howard came to Berkeley, directing the architecture school from 1903 through 1928 and implementing a much modified version of the Bénard plan.

While teaching and supervising development of the university, Howard also maintained a limited private architectural practice, with most of his later work clearly showing the classical influence of McKim, Mead & White and the Ecole des Beaux Arts. His earliest private work, however, was, as Sally Woodbridge describes it, "Craftsman and woodsy." Good examples are three houses he designed in 1904 for the Frances Gregory family in the Berkeley hills. Better still is the house the Gregorys financed for Howard in 1912.

The house is situated on a corner lot with magnificent views of the bay, and to take advantage of these Howard bent the house plan around two polygonal bays that accommodate the living room and the original library. To the rear of the house, these wings enclose an intimate patio at the bottom of a large hillside garden. Oversized shingles sheath the entire house, laid in a distinctive pattern with one band showing nearly its full length, but the one underneath showing only about two and a half inches. This results in strong horizontal lines running across the facades. Offsetting the large scale of the shingling are the small diamond panes in the windows.

When the house was sold in the 1920s to the chairman of the university's English department, his book collection necessitated the addition of a larger library. Adapted to the shape of the site and adding a new bend to the plan, the new library wing was designed by Julia Morgan. With her characteristic sensitivity to place and context, she maintained the external shingle pattern, and the new wing seems to have always been part of the house. It is joined to the dining room wing of the older house by a shingled tower containing a spiral staircase, a lovely counterpoint to Howard's blunter enclosure of the main staircase.

John Galen Howard's house demonstrates that, even in the hands of an avowed classicist, the shingle tradition could continue to persuasively evoke domesticity and rustic comfort well into the twentieth century.

JOHN S. THOMAS HOUSE

Berkeley, California, 1914

The pioneering work of the first generation of architects to employ the Shingle Style in the San Francisco Bay area was soon followed by that of a second generation of architects who continued this local shingle tradition, branching out in new directions. One of these was William Charles Hays, who trained in the office of John Galen Howard, the planner and builder of the campus of the University of California. Hays' John S. Thomas house of 1914 demonstrates this connection to Howard. From his training in the office of McKim, Mead & White, Howard had acquired somewhat didactic preferences for formality and balance, at least in public and institutional buildings. But in his own Berkeley residence Howard worked in a more relaxed way, wrapping tightly-organized spaces in an abstracted picturesque informality derived from the Shingle Style. In the Thomas house, Hays effected his own reconciliation of these seemingly opposite inclinations.

Howard's residence was built a few blocks north of the Berkeley campus, on LeRoy Avenue, where the land begins to rise sharply into the Berkeley hills.

RIGHT: *Once a sleeping porch, the upper loggia looks to the far reaches of San Francisco Bay.*

OPPOSITE: *A free classicism organizes this shingled facade.*

A block further to the east Buena Vista Way begins its serpentine climb, and here Hays built the Thomas house. The carefully composed symmetry of this house, with its low, broadly extended roof, has just a hint of Otto Wagner's own classical suburban villa outside of Vienna (built 1886-88). Both sit high on hillsides, recessed in lush plantings like garden pavilions, and both have upper loggias open to the view. But Wagner made direct reference to the classical formality of the Italian Renaissance architect Andrea Palladio, whereas Hays created a rustic city house, using overscaled redwood shakes. In the otherwise crisply rectilinear facade, Hays introduced an arched recessed entry. This arch makes the viewer aware that despite his hints of classicizing order, Hays has devised an organizing system of seven posts defining six bays, whereas conventions of classical design would have insisted on eight supports defining seven bays, with the entry in the central bay. Instead, as the present owner of the house likes to jest, the entry is placed in a location well suited to Berkeley—somewhat left of center.

In the light of a golden evening, the shingles of the Thomas house take on the color of old burnished copper, and provide an appropriate ending to this phase of development of the Bay Area shingle style.

GUY HYDE CHICK HOUSE

Oakland, California, 1914

The unusual qualities noted in Bernard Maybeck's Hopps house, Grayoaks (1906), were merely the overture to even more delightfully playful combinations of materials and details he would use in an Oakland hillside house for the Guy Hyde Chick family, 1913–14. Like the Hopps house, placed with its longer dimension crosswise on the hillside, the Chick house is loosely inspired by Swiss chalets, with a broad, low-pitched roof extending well over the walls.

The walls are sheathed largely in shingles, with the cantilevered sections of the upper floor enclosed by vertical board siding punctuated with battens of varying size, two slender battens for each larger batten. At the base of the lower walls, the exposed edges of the shingles are cut into a fine sawtooth pattern that forms a distinctive decorative band. A similarly fine sawtooth band runs over the windows. The courses of shingles over the Tudor-arched entry openings are cinched up over the entrances, like a curtain being ever so slightly raised; they gently rise and fall, maintaining their continuous flowing lineaments over the openings.

RIGHT: *The trellis over the front door supports a venerable wisteria.*

OPPOSITE: *The stair hall.*

Enormous windows merge the gardens with the redwood paneled interiors.

Maybeck's inventive whimsy can be seen both in the heavy medieval quatrefoil balustrade of the balcony that extends from the midfloor landing of the stair, and in the lively paint colors used on the wood surfaces—deep blue and yellow at the recessed entry, red on the exposed rafters of the eaves. The undersides of the eaves were painted yellow, reflecting warm light into the otherwise shadowy bedrooms on the upper floor. Additional indications of Maybeck's adroit playfulness are the trellises at the end of the roofs and over the the front door. The branching supports of the entrance trellis echo the sturdy old oak trees that ring the house, and the open frameworks of the trellises mimic the leafy canopies of the trees and the dappled light they cast upon the house.

On the upper floor, one of the rooms was originally a screened sleeping porch (since glazed and enclosed). In this room, closely linked with the out-of-doors, Maybeck used a deliberately rustic treatment, finishing the inner surfaces with rough-sawn board and batten siding, exposing the paired roof rafters, and carrying them on heavy purlin beams braced by scroll-sawn brackets. All the construction is revealed, even to leaving the stamp of the Stege Lumber Co. on one of the purlins. Maybeck's interest in the direct expression of construction materials is also seen in the exposed concrete used in the base walls of the house, in the chimneys, and in the curving seat incorporated into the trellis support at the front door.

The recessed lower level entrance opens to a broad hall, with the spacious living room to the left and the dining room to the right. The hall and main rooms are paneled in redwood, whose color imparts a warmth and intimacy to these spacious volumes. Here and there, repetitions of the Tudor arch

OPPOSITE: *The concrete chimney dominates the west end of the house. Latticed overhangs complement the canopies of the surrounding oak trees.*

The library fireplace.

OPPOSITE: *A gothic arch allows a narrow doorway to fit under the staircase.*

form are used, and the post at the foot of the stair is carved in the form of a gigantic urn. Contrasting with this eclecticism are elements—such as the spare paneling, the expanses of glass, and particularly the austere concrete grid of the library fireplace—that anticipate modernist developments of the next half century. It was such crisp minimalism that led Bay Area modernists of the 1950s to see the elderly Maybeck as a kind of professional godfather.

But few architects since Maybeck have made building and living in a house such fun. Maybeck's creativity contained elements of perception and wonder more common in children than adults; his residences are amusing without ever threatening to become hollow jokes. The shingles he used so often in his houses make them part of the extended heritage of the Shingle Style. And in his attention to the natural qualities of materials, in his delight in shapes and colors, in his juxtapositions of the familiar and the exotic, Maybeck's houses demonstrate that he is fittingly viewed as a latter-day master of the Arts and Crafts movement, for he truly made "Art that is Life."

SAUSALITO WOMAN'S CLUB

Sausalito, California, 1917

In the early twentieth century, women's clubs sprang up across America as public-spirited women, denied their power at the ballot box, created their own places to organize and effect social change. Here they addressed important issues of the day, such as temperance, child labor, and women's rights.

These clubs, and the urban settlement houses run by women, comprised what amounted to the Women's Network, as writer Sara Holmes Boutelle has dubbed it. Many female architects and designers proved extremely sensitive to the needs of women's organizations, including Hazel Wood Waterman of San Diego, her protégé Lilian Rice, Minerva Park Nichols, and Louise Blanchard Bethune.

Perhaps the most prominent of these architects was Julia Morgan of San Francisco, who designed a wide range of buildings for women's service organizations. Today Morgan is most commonly remembered as the designer of William Randolph Hearst's San Simeon, a grandiose smorgasbord of Spanish and baroque architectural fragments that rose on the California coast from 1919 to 1951. The Sausalito Woman's Club is a fine example of Morgan's

RIGHT: *Like many of Julia Morgan's shingled buildings, the club house combines beautifully effective design with self-effacing modesty.*

OPPOSITE: *A painting celebrates the view from the heights of Sausalito.*

ability to create, with only a minute fraction of the San Simeon budget, a building of far greater vigor and originality.

Morgan (1872–1957), a San Francisco native, was unusual in her profession in many ways. She was one of the first American women determined to become an architect. She earned her baccalaureate degree in engineering at the University of California at Berkeley and she finished her architectural education at the prestigious Ecole des Beaux Arts in Paris. She was the first woman admitted to the architecture program at the Ecole, largely because officials never anticipated a female applicant and had established no rules forbidding it. After four years of study in Paris, she returned to San Francisco, working in the office of John Galen Howard and re-establishing her close professional association with Bernard Maybeck.

In many of her early residential commissions, Morgan worked in the shingled Arts and Crafts idiom being used by Maybeck and other Bay Area architects. This can be seen in her remarkable house for Helen Livermore on Russian Hill in San Francisco. This cubic massing of interlocked shingled blocks was built in 1916–17. Morgan would use a similar Arts and Crafts simplicity of construction and detail in the club house she designed simultaneously at Sausalito.

The Sausalito Woman's Club emerged from the civic concern—if not outrage—that spurred many women into action. Old-time members recalled the story well. In 1911 Mrs. R. W. Wood was alarmed to find workers busy preparing to fell a row of mature cypress trees along Bulkley Avenue. She was told that the town had just replaced its horse-drawn taxi with a model-T Ford and the road had to be regraded to accommodate the automobile. Hurrying back up the hill to Harrison Avenue, she recruited neighboring wives to force a judge to issue a restraining order. By the time they returned to Bulkley Avenue, one lone cypress remained, but it was saved. It lived out its natural life span, becoming a massive and towering tree. For decades cars had to circle around what was affectionately called "The Founding Tree."

With that success under their collective belt, the women of Sausalito were frustrated that the town's male Board of Trustees ignored their requests and admonitions. On March 28, 1913, they formed the Sausalito Woman's Club and organized four divisions: Civic, Outdoor Art, Music, and Literary. The mission of the Woman's Club was "to preserve the beauty of Sausalito and protect its best interests." The group met in rented quarters. They recruited architect William A. Faville to design a fountain and sculpture for the town park, and they also obtained permission for Faville to supervise the proposed removal of further trees. When the United States became involved in World War I, club women engaged in preparedness activities and packed medical supplies. In the next several decades the Sausalito Woman's Club would help

Even at the back of the stage, french doors open to a view from a balcony.

189

launch the local Chamber of Commerce and the Lions' Club, lead efforts to keep fish canneries off the Sausalito waterfront, and preserve significant municipal landscapes. Members also drafted a zoning ordinance adapted and enacted by city officials. During the Depression, the club conducted fund collections and clothing drives, and during World War II members again worked in many support activities.

In 1917, Mr. F. A. Robbins presented the club with property at 120 Central Avenue and the funds to build permanent headquarters in memory of his wife, Grace McGregor Robbins, who had been closely associated with the club's founders. It was singularly appropriate that the women then turned to a woman architect to design their club house, and especially appropriate for them to select Julia Morgan, who had already designed a number of women's clubs and had begun a series of YWCA buildings. With a limit to funds available, and perhaps due to wartime stringencies, Morgan designed a compact and lean building of redwood, sheathed with redwood shingles.

The window-wall system shows the care Morgan took with this design. The French doors that open onto the terrace are proportioned so that their

The upper floor of the club holds dressing rooms for amateur theatricals, with sweeping views of San Francisco Bay.

width to height is nearly the golden section, and the ratios of width to height of the individual lights in the windows, as well as the entire bank of French doors and transoms, form a series of complementary relationships. Moreover, the thickness of the French door frames is contrasted to exquisitely thin muntin strips that hold the glass panels. The lightness of this glazed wall system is reinforced by the delicacy with which the shallow triangular roof trusses are themselves carried by transverse trusses across the principal room. The diagonal bracing members are all slightly curved to break what might otherwise have been a harsh and overly insistent geometrical system.

Morgan's simple building took good advantage of the hillside site, providing sweeping views of San Francisco across the bay. Window-walls of ranked French doors with transoms, stretch from floor to ceiling on two sides, with balconies and terraces opening the building everywhere to the outside, to the views of a landscape that the club has worked to preserve and protect for most of the twentieth century.

TIMBERLINE LODGE

Mount Hood, Oregon, 1936–38

A ram smiles from the huge knocker on the lobby door.

OPPOSITE: *This icon of Northwest regional architecture incorporates many characteristics of the Shingle Style: encompassing roofs, unified window groupings, and the direct use of rustic materials.*

Rising to 11,245 feet, Mount Hood can be seen from nearly one hundred miles in every direction. In Portland, which lies sixty-one miles west of the mountain, the free-standing white conical peak of this dormant volcano glimmers pink in the alpenglow of long summer sunsets. When all-weather roads began to reach its southern slopes, Portlanders were drawn to the mountain for winter athletic activities, especially skiing and winter sports festivals. In the twenties, as skiing increased in popularity, it became evident that an all-weather facility was needed on the mountain to welcome sports enthusiasts and to provide meals and lodging. The drive back to Portland was just long and difficult enough to encourage overnight stays. Such a public facility would also provide an important jumping-off point for hiking on the mountain during summer months.

The placement of such a lodge was the subject of vigorous debate. Some planners felt it should lie near the state highway that curved around the south flank of the mountain, at an altitude of about five thousand feet. Others, including the young Portland designer John Yeon, an avid skier, successfully argued that the lodge should be built a thousand feet higher, at the timberline. At that higher elevation, a mechanical tram would be unnecessary; skiers would have direct access to slopes covered with deep winter snow and unobstructed by trees. In 1929, just as discussion pointed toward construction of a lodge at the timberline, the beginning of the Great Depression squelched any idea that the city or private groups might carry this project to fruition.

When work relief programs were begun under President Franklin D. Roosevelt, the idea soon emerged of making construction of the Timberline Lodge a federally sponsored project, under the direction of the United States Forest Service. It was decided that the entire lodge would be a public works project, utilizing extensive handcraftsmanship, providing work for wood carvers, carpenters, weavers, seamstresses, metal-workers, and painters and other artists who would be commissioned to embellish the building inside and out. The interior design was done by Howard Gifford, who developed the abstract decorative motifs used throughout the building. They were purportedly based on Native American designs he found in his daughter's Campfire Girls handbook (the handbook today is on display in the lodge). Production of the artwork, furniture, and other artistic fixtures was directed

by Margery Hoffman Smith, a Portland interior designer active in the local Arts and Crafts movement who was appointed supervisor of the Oregon Art Project of the WPA.

The building consists of two major wings, the longer one housing hotel rooms, and the shorter one accommodating restaurant, kitchen, snack bar, and ski-rental facilities. (In 1975 convention facilities were appended to the kitchen area.) The two wings connect at an angle of 120 degrees to a large hexagonal central tower. Inside this tower is a lobby of heroic size and imagery. Encircling balconies and staircases make shadowy enclosures around a towering central chimney mass nearly four stories high; fires blaze at its base in two back-to-back stone fireplaces of walk-in dimensions.

At this altitude, where snow depths often exceed twenty feet, it was essential to design the building to withstand tremendous weight; this engineering was done by W. D. Smith and W. W. Gano. To counter lateral creep of the building from the pressure of snow, the foundations and lower walls of the building are massive formations of concrete, faced with large basalt boulders, graduated in size and suggesting walls of solid masonry. Heavy snow also encouraged the use of long, steep roof planes covered with long wooden

In the dining room, massive knee-braces articulate the window-wall.

OPPOSITE: *Mountain wildlife adorn the dining room fireplace. The chair displays the "Timberline arch" motif used throughout the lodge.*

OVERLEAF: *Steep roofs create a fitting resemblance to the surrounding mountains.*

shakes one-and-a-quarter inches thick at their base and laid three deep at the eaves. When winter snow filled the spaces beneath the long roof slopes, some early skiers climbed up to the chimney top to glide down the roof to the snow fields below. The vertical walls are sheathed in vertical board and batten siding. These walls, along with the shingles, were originally left to weather naturally to a silver sheen that complemented the grayish brown of the rock-faced walls below.

Though intimate in scale, Timberline Lodge is built of tremendously oversized structural elements. The irregular surfaces and natural materials make it apparent that this is one of the last great handmade public buildings, combining Art Deco forms with Arts and Crafts attention to detail. Treasured by citizens of the region, it has been lovingly restored—the original worn and faded appliqué coverlets, draperies, and threadbare hand-woven rugs have been replaced by new textiles sewn and woven in the original patterns by volunteer craftspeople.

Timberline Lodge was instrumental in giving continued life to a Northwest regional tradition of boldly designed, shingle-clad buildings, exploiting wood in structure and finish. This regionalism, grandchild of the Shingle Style, would re-emerge in the late 1930s in the architecture of Pietro Belluschi and John Yeon and, again, reinvigorated, in the 1960s and 1970s, re-establishing a tradition of wood and shingle building that continues today.

THE FOUREST

Kentwoodlands, California, 1957

In the 1950s and 1960s, in and around San Francisco, the local architectural tradition known as Bay Area Regionalism was reinvigorated. Buildings by architects such as Coxhead, Polk, Maybeck, Morgan, and Howard were examined and appreciated anew. The lineage of the eastern Shingle Style and the Bay Area shingled tradition has been continued in many of the private residences designed by Joseph Esherick. It is singularly appropriate that Esherick should create this link, for he drew on a profound family tradition that sought to unite art and craft. Born in Philadelphia in 1914, Joseph Esherick was the nephew of Wharton Esherick, a major artistic presence in Philadelphia in the 1930s and 1940s.

Wharton Esherick spurned the conventional modernist emphasis on metal and glass, turning instead to a fuller realization of the artistic potential of wood. Furniture design was one of Wharton Esherick's special interests, and his early work stressed strong angular geometries, softened by the grain and color of the beautiful woods he used; his later pieces were more fluid in line and hence emphasized even more the organic nature of the construction material.

RIGHT: *The house is organized under a broad triangular roof.*

OPPOSITE: *Like a traditional porch, the pergola was designed to shade the patio and principal rooms of the house.*

The stair hall offers carefully framed
glimpses of the surrounding rooms.

The upper stair landing overlooks the living room hearth.

OPPOSITE: *Simple materials are joined with exquisite care.*

Joseph Esherick studied architecture at the University of Pennsylvania, graduating in 1937, at a time when the formal influence of Paul Philippe Cret was still strong but the modernism of George Howe was beginning to assert itself. Esherick, as a result, would work toward fusing this interest in building form with the expression of modern materials. After a short summer visit to San Francisco, Esherick decided to settle there, working in the office of Walter Steilberg, the former head draftsman and engineer for Julia Morgan. It was through Steilberg that Esherick learned of Morgan's early residential work and her use of wood and shingles. He also met the aged Maybeck, studied his architecture, and was introduced to the shingled work of Polk by William Wurster, who had a small apartment in Polk's Russian Hill house.

After serving in the navy during World War II, Esherick set up his own practice. He designed small houses and apartment complexes in and around San Francisco, exploring spatial complexities and modular clarity and using wood siding. In 1950 he and his wife designed a house for themselves in Marin County. The major rooms were placed in an extended rectangular block capped with a long, low gable roof, recalling the broad triangular roof of McKim, Mead & White's Low house. Inside and out, vertical wood siding or paneling was employed. In a landscape designed by Lawrence Halprin, the Esherick house is turned so that the extensive banks of glass in its double-height living room face east and are sheltered by the canopy of a magnificent old oak tree.

In 1957 Esherick returned to these themes in the design of The Fourest, a house for a dentist and his family in nearby Kentwoodlands. The name is a play on words, alluding both to the address of the house and to the four stunningly beautiful daughters who grew up in it. Facing southwest in a landscape by Richard Heims, The Fourest is again a broad triangle, this time sheathed in shingles, with the great living room set on the southerly side of the long rectangle, and the somewhat more private family spaces of den and library placed in the northerly end. Within the broad shingled southwest facade, the windows are grouped with relaxed informality according to the needs of the interior rooms. A free-standing pergola just beyond the long garden facade shades the glass of the living and dining areas. The pergola creates a sprinkling of dappled light over the outdoor patio (or piazza, to use the term associated with the eastern Shingle Style houses). Inside the house, vertical wood paneling echoes the warm colors of the shingled exterior. In keeping with the architectural orthodoxies of the 1950s, the only ornament present in this house is derived from the carefully considered and elegantly resolved intersections of structural elements. In this fondness for joinery, Esherick may have been inspired by the houses of Greene & Greene, which also were being rediscovered and appreciated anew in the 1950s.

FLINN HOUSE

East Hampton, New York, 1978–79

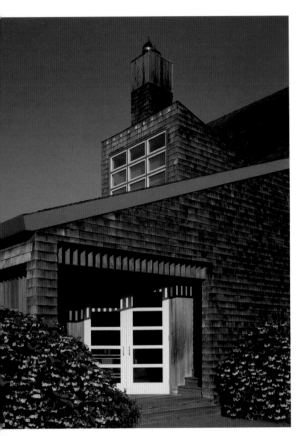

Jaquelin Robertson abstracted elements of colonial salt-box houses, which had also inspired McKim, Mead & White.

OPPOSITE: *Inspiration was also found in the latticework of Shingle Style houses.*

OVERLEAF: *The house faces a broad lawn, and the ocean just beyond.*

Although Robert A. M. Stern is most commonly identified with the resurgence of the Shingle Style in the 1970s, especially in those towns on the South Fork of Long Island that are collectively known as the Hamptons, other architects were also busy with their own explorations of the shingled house. An excellent example of such work is Jaquelin Robertson's Flinn house in East Hampton. Perhaps more than any other, this compact house recalls the ancestral New England saltbox house, with its long, sloping, bent roof extending to the rear. In fact, Robertson had in mind a specific 1735 saltbox house, but he exaggerated the steep slope of the main upper roof, while extending even more the slope of the lower roof. It is intriguing to compare the rambling and relaxed collection of roof planes and masses at the rear of the Flinn house with the rear of the Bishop George Berkeley house, which had so inspired McKim a century earlier. Both designs playfully exploit random but purposeful arrangement of elements.

When Robertson was engaged to design the Flinn house in 1978, he had already been frequenting East Hampton for eight years, having remodeled a house there for his own family in 1970. His knowledge of the local architecture character was firsthand.

Shingled farm houses on Long Island were originally built without extensions and with small, tight windows, partly due to the expense of glass but also to keep out winter winds. Nowadays concerns are different, as reflected in Robertson's design, with the projecting rear wing sheltering an open court. Robertson gives an explanation of the design: "Formally the house is a response to the general 'manners' of an historic summer residential colony with a distinctive architectural tradition—and a demanding climate. The archetypal local dwelling is the New England salt box amended over the years by the 'amenities' of a more easy-going summer cottage style—porches, rooms-as-bay-windows, inglenooks, etc.—and the house has consciously attempted to employ these images so as to fit into a popular, genteel, and still valid visual and social environment. Materials, massing, scale of openings, trim color, roof silhouette and siting with respect to lot lines and other buildings are within an established local vocabulary."

Robertson opened up the house with large windows to the view, flooding the interior with light. As he notes in an interview in the December 1980

The living room, with its updated ramma beam and latticework, is derived from the interiors of older Shingle Style houses.

OPPOSITE: *The entry hall, with an upper overlook from the living room inglenook.*

issue of *Architectural Digest*, "The key to all houses, in fact the key to all architecture, is light.... Very consciously this house was designed to maximize the way in which light changes spaces and produces patterns." One way that Robertson produced visual and light patterns was through extensive use of latticework, just as in the original Shingle Style interiors. As in McKim, Mead & White's Newcomb house, a continuous *ramma* beam runs around the upper level of the living room; above this at intervals are latticed transom screens. Elsewhere in the upper regions, large lattice screens provide privacy and facilitate natural ventilation.

In many details, the house was designed to accommodate the lives of the owners and their three growing children. One example: external staircases lead directly to the upper level from both the entry porch and the seaside front of the house, so that after an afternoon on the beach or golf course, the parents could easily retreat to their own quarters to freshen up for the evening, without colliding with the children and their guests on the lower level.

Not nearly as well known as examples by other architects, the Flinn house is among the most authentic descendants of the original Shingle Style in its directness, responsiveness to internal functions, and playful manipulations of space and building masses. Nowhere does it descend to parody or easy imitation. Like the Shingle Style houses that preceded it by a century, the Flinn house promotes an informal life by making a strong connection to the surrounding landscape and the sea.

LAWSON HOUSE

East Quogue, New York, 1979–81

A curvaceous staircase makes an event of every arrival and departure.

OPPOSITE: *The house was designed as a square pavillion atop a sand dune.*

Robert A. M. Stern was among the first of his generation of architects to look seriously at the Shingle Style of the 1880s and 1890s as an inspiration for contemporary design. Perhaps this was inevitable, given his years of study at Yale and the persuasive influence of his teacher Vincent Scully, who had resurrected and christened this phase of American architecture.

Stern began drawing on Shingle Style prototypes in his first independent commission, the Wiseman house built in Montauk, Long Island, New York, in 1965–67. Montauk is the site of a grouping of six shingled houses that McKim, Mead & White designed in the 1880s for a cooperative association of New York businessmen. The houses of the Montauk Association still stand, near the very tip of Long Island. Although not well known, they form an important ensemble of buildings.

Stern's Wiseman house, a rather free-form block that refers to the broad triangle gable motif, is completely covered in shingles. Stern designed a larger shingled house at Montauk in 1971–72, another in East Hampton in 1973–75, and a more formal year-round shingled house in 1976–77 on Mount Desert Island, Maine. In 1978 he was engaged to make an addition to the end of a large Shingle Style house in Dublin, New Hampshire, designed by Peabody & Stearns in 1887; his addition merges nearly seamlessly with the original house. With such projects, Stern established himself as one of the most knowledgeable and respectful of the new architects working in the revived Shingle Style idiom.

Stern's concern for maintaining continuity of character led shortly to a commission for renovating a shingled East Hampton summer cottage of 1891 by architect I. H. Green, modifying it for year-round living, and opening it to better light. Existing window casing and columns were saved and relocated, resulting in what editors Arnell and Bickford described as a "sympathetic and scholarly conversation between the present and the past."

The emergence of Stern's large formal shingled vacation houses was marked by his house at Chilmark, Martha's Vineyard, in 1979–83, where his restrained low-slung composition, with commodious wraparound porches and shingled covering, blends well with the vintage shingled buildings on the island. Immediately after this Stern designed another large shingled house on Martha's Vineyard, at Farm Neck, with its living spaces pulled within a broad

The barrel-vaulted window lends the master suite spatial drama to match the ocean view.

archetypal triangular gable, paying homage to McKim, Mead & White's Low house and perhaps to Schweinfurth's Unitarian Church in Berkeley.

From this point on, the shingled houses by Stern and his associates expanded in size, and gradually grew more derivative in their detailing.

In this progression of shingled summer houses, the compact and pavilion-like Lawson house, built in 1979–81 on the southern Atlantic coast of Long Island, at East Quogue, takes on particular importance, for it is among the most succinct of Stern's shingled designs. It retains the inventiveness of Stern's earliest shingled houses, with a pure simplicity of form, while incorporating historicist details that become important accents rather than components of a conscious revival.

Stern formed the Lawson house as an iconic pavilion, a building reduced to its essentials. Because the house was placed on the rise of a sand dune, it was possible to place three compact guest bedrooms on the lower level, with windows to the north, while keeping the main floor of the house open on the upper level. The principal bedroom is located in the gable, with a broad, hugely scaled, semicircular window looking south to the sea, and a broad eyebrow dormer on the north side lighting the master bathroom. The broad and slightly flared hip roof rises in a central gable section over the bedroom. Windows on either side provide cross-ventilation.

On the main floor, a simple and straightforward structural logic is expressed by the grid of white Tuscan Doric columns, some inside the house, some on the open porches, and one (at the corner of the kitchen) expressed externally by pulling in the shingled wall behind it where the kitchen cabinets meet. Like Wright's small houses, the plan is pinned to the earth by a central chimney mass. The living room and its flanking subspaces have an axial, almost formal character, which was relaxed by the canted wall on the northwest corner with its recessed, built-in seat. As Peter Arnell, Ted Bickford, and Luis F. Rueda note in their monographs on Stern's work, the Lawson house deliberately drew on a regional variant of the Shingle Style, particularly the kind of small beach cottages that sprang up along the East Coast in the 1910s and '20s. These cottages' astylar simplicity and direct use of materials, they suggest, were influenced by the writings of Gustav Stickley.

PETRIE HOUSE

Wainscott, New York, 1982

A massive mahogany bench on the back porch overlooks the ocean.

OPPOSITE: *Fences echo the rolling surf.*

In the 1970s, the Hamptons on the southern shore of New York's Long Island became a proving ground for attention-getting statement houses by ambitious architects. Some were slant-roofed, discordant assemblages of boxes; others followed an emphatically historicist and increasingly popular Postmodern Shingle Style. When Donald Petrie purchased a parcel of land on a salt pond behind the ocean dunes in the old town of Wainscott, he wanted something unusual for the fashion-conscious Hamptons: a house that would reflect and defer to the unpretentious turn-of-the-century shingle and clapboard houses of his neighbors. Petrie put up a scaffold on his uncleared land to locate the best ocean view. Drawing on his experience as an accomplished sailor, he studied the prevailing winds, for he wanted the proposed house to cool itself naturally.

After examining the work of several architects, Petrie turned to the Philadelphia firm of Venturi, Rauch & Scott Brown. He knew of Robert Venturi's paired Trubeck-Wislocki shingled cottages on Nantucket (1970) and the Coxe-Hayden studio houses on Block Island (1979). Petrie also admired Venturi's design of a larger house in Tucker's Town, Bermuda (1975), because it responded to an oceanfront site while incorporating references to traditional Bermuda architecture. Petrie desired such a house, responsive to the unique qualities of his site and to local architectural traditions, especially to the sort of shingled cottage with gambrel roof that Stanford White designed in 1891 for his painter friend William Merritt Chase in the nearby Shinnecock Hills.

The house designed by Venturi for Petrie is similarly covered by a shingled gambrel roof, with large cantilevered eaves to shelter the long porches beneath. This roof form is much like that favored by the first Dutch Walloon settlers on Long Island. Venturi also incorporated long shed-roofed dormers, with windows irregularly placed here and there according to internal needs. Venturi described his intent: "The building is purposely understated, and then there are touches that show you that we *mean* the understatement—for example, the use of painted panels on the dormers. The waves symbolize the nearby sea and the owner's love of sailing; on another level, this ornamentation is saying, 'Here is not a wall but a window that was left out.'" Venturi also had in mind similar long dormers on Shingle Style houses designed by

The house possesses the relaxed inevitability of its older neighbors.

OPPOSITE: *Robert Venturi updated the gaily decorated dormers of early Shingle Style houses with a touch of Pop Art.*

OVERLEAF: *The living room.*

Henry Hobson Richardson and McKim, Mead & White that included ornamental bas-relief stucco panels between the windows. This abstracted historicism is accentuated by Venturi's use of ordinary sheet metal chimneys in place of traditional brick. Similarly, posts and railings are banished from the porches; overscaled mahogany benches anchor the ends of the back porch.

Well known for his preference for "complexity and contradiction" in contemporary architecture, Venturi cultivated these qualities in the Petrie house. Although the house itself appears rather formal, it has a loose and informal relationship to its adjoining outbuilding, a combined pool house, garage, and sail loft. The compact formal garden contrasts to the wild meadow and woods nearby. And there are also the large, overscaled details, such as the big dentil molding and chair rail in the living room. As Venturi noted, "ordinary elements become extraordinary when you do something 'wrong' with them. In this way, we exemplify old traditional house details."

The main floor is arranged with informal practicality. An ample eat-in kitchen is adjacent to a large multipurpose social room. This room, akin to the living halls of older Shingle Style houses, measures 23 by 36 feet, with screened French doors on either side opening to the sheltered porches and providing the requisite cross-ventilation. The upper floor is also simple in

In the pool house.

plan; the bedrooms and a study are strung along a corridor, so that all the rooms face the water. Some open onto a balcony cut into the gambrel, affording views of the treetops, the pond, and the ocean.

What can't be seen, however, is the sturdy construction of the house and its massive foundations, designed to withstand gale-force winds. With its long side turned to the sea, the pent roof of the house can act as an air foil. One of the builders observed that he would remember this house when another 1938-style hurricane hits Long Island, "because this is where I want to weather it." Great care was also taken with the refined detailing of the house, with its craftsmanship and its exceptionally high degree of finish. Perhaps the builder had this in mind, too, when planning to weather a storm here, knowing that in every way the Petrie house is a thing well made.

KRAGSYDE

Swan's Island, Maine, 1982–

> *Make no little plans; they have no magic to stir men's blood.... Make big plans; aim high in hope and work, remembering that a noble, logical diagram once recorded will never die, but long after we are gone will be a living thing, asserting itself with ever-growing insistency.*
>
> —DANIEL BURNHAM

The luggage door under the great arch.

OPPOSITE: *A faithful re-creation of Peabody & Stearns' lost masterpiece; a labor of love rather than of architectural theory.*

Burnham's words were originally framed as a challenge to early urban planners at the dawn of the twentieth century, but they have special relevance to the Goodrich-Beyor house. They speak of the power of a wondrous design that will not let itself be quelled, that asserts itself and cries out for realization. So it has proven to be with perhaps the most picturesque and engaging of all the original Shingle Style houses: Kragsyde, built for G. Nixon Black at Manchester-by-the-Sea, Massachusetts, in 1882–84, from designs by the Boston architects Peabody & Stearns. The house seemed part of its rugged coastal landscape, its great Richardsonian arch, wood-framed and shingle-wrapped, rising to cover the porte-cochère. It was a spacious house, with over thirteen rooms. But the family occupied it for only a few summers before Black died. It passed to a younger generation, and, by 1929, it had been pulled down.

Such was the power of Kragsyde, rediscovered by Vincent Scully in *The Shingle Style*, that when Jane Goodrich looked through the book as a child she was captivated by it. She read the book again in college and decided she wanted to live in that house. Later, she and her husband, Jim Beyor, visited Manchester to see the masterpiece for themselves. They discovered it was long gone, surviving only as a small black and white drawing in a double-page spread of vignettes about the house. They resolved they would undertake nothing less than the reconstruction of the house on a new site that still possessed the natural quiet once found on the north shore of Massachusetts. Their reconstruction would rise on a craggy granite outcrop on the shores of Swan's Island, Maine.

Beyor, a builder by trade, and Goodrich, a graphic designer and publisher, "were moved by the rambling, gentle, carefree life" embodied by the original house and were convinced that "this house was worthy of being rebuilt," as they related in *Progressive Architecture*. Unable to put the house back on its original site, they looked for another, as wooded and as similar to the

A screened porch off the dining room offers tree-top views of the ocean.

OPPOSITE: *In the living hall, the fireplace nook displays souvenirs of journeys both real and imagined.*

OVERLEAF: *Peabody and Stearns' boudoir with its wrap-around porch now serves as a library for the lady of the house.*

original as possible, eventually settling on a six-acre parcel on Swan's Island, Maine. Because of the particular topography of the island they decided to invert the plan, building a mirror image of the original that is better fitted to the rock forms on their site. The couple found the original drawings in the Boston Public Library, prepared their adaptation, and began work in 1982, exactly a century after the original Kragsyde was built.

Work has progressed slowly, being done mostly by Beyor and Goodrich themselves. Traditional construction methods have been employed, the heavy kneebraces being hand-cut, the pieces lifted into place by block and tackle, with modern industrial materials like plywood being shunned. The nearly two hundred windows in the house were handcrafted by Beyor over the course of a year. Inside the house Beyor and Goodrich have made only small concessions to life at the end of the twentieth century. "We're Luddites out here," laughs Goodrich. The kitchen has been made larger and more serviceable than that of the first Kragsyde, and a room noted as a bedroom on the original plans is now a bathroom. The room over the great arch is a library instead of a boudoir. Little else is different.

Some might quibble over the resurrection of someone else's design made for another place, another time, but for those who wish to recapture the feeling of another time, to truly live in it, these types of houses and interiors need to be re-created. Far too much of our national cultural memory has been obliterated, often without comment or notice, for it is lost slowly, morsel by morsel, building by building, just a little at a time. But the aggregate loss is stupefying. Perhaps if dedicated craftspeople are willing to take on the discipline of remaking a lost masterwork on its own terms, such literal re-creation is right. It becomes an affirmation of thoughts and artifacts we still hold important. Too often it is not until after the loss that we realize how much we value what is gone. At Kragsyde, we have a second chance to see a wonderful house.

226

The staircase at Felsted.

BIBLIOGRAPHY

Arnell, Peter, and Ted Bickford. *Robert A. M. Stern, 1965–1980: Toward a Modern Architecture After Modernism* (New York: Rizzoli, 1981).

Beach, John. "The Bay Area Tradition, 1890–1918," in S. Woodbridge, ed., *Bay Area Houses* (Salt Lake City: Peregrine Smith Books, 1988).

Bosker, Gideon, et al. *Timberline Lodge: A Love Story* (Portland, OR: Friends of Timberline, 1986).

Bosley, Edward, R. *Gamble House: Greene and Greene* (London: Phaidon, 1992).

Boutelle, Sara Holmes. *Julia Morgan, Architect* (New York: Abbeville Press, 1988).

Brunk, Thomas. "The House That Freer Built," *Dichotomy* (University of Detroit School of Architecture) 3 (Spring 1981): 5–53.

Cardwell, Kenneth H. *Bernard Maybeck: Artisan, Architect, Artist* (Salt Lake City: Peregrine Smith Books, 1983).

Coxhead, Ernest. "Church Planning," *Architectural News* (November 1890): 5–7; (December 1890): 19–20; (January 1891): 25–26.

Creese, Walter L. *The Crowning of the American Landscape: Eight Great Spaces and Their Buildings* (Princeton: Princeton University Press, 1985).

Current, William R., and Karen Current. *Greene & Greene, Architects in the Residential Style* (Forth Worth: Amon Carter Museum of Western Art, 1974).

Downing, Antoinette F., and Vincent J. Scully, Jr. *The Architectural Heritage of Newport, Rhode Island*, 2d ed. (New York: Clarkson N. Potter, 1967).

Draper, Joan. "The Ecole des Beaux Arts and the Architectural Profession in the United States: The Case of John Galen Howard," in Spiro Kostof, ed., *The Architect: Chapters in the History of the Profession* (New York: Oxford University Press, 1977).

Elliott, Maud Howe. *This Was My Newport* (Cambridge, MA: Mythology Company, A. Marshall Jones, 1944).

Fahlman, Betsy. "Wilson Eyre in Detroit: The Charles Lang Freer House," *Winterthur Portfolio* 15 (Autumn 1980): 258–59.

Ferry, W. Hawkins. *The Buildings of Detroit: A History*, rev. ed. (Detroit: Wayne State University Press, 1980).

Floyd, Margaret Henderson. *Henry Hobson Richardson: A Genius for Architecture* (New York: Monacelli Press, 1997).

Gebhard, David, ed. *The Work of Purcell and Elmslie, Architects* (Park Forest, IL: Prairie School Press, 1965), a collection of three articles on the work of these architects published in *The Western Architect*, January 1913, January 1915, and July 1915.

Gill, Brendan. "Kingscote," *Architectural Digest* 48 (October 1991): 30–34.

———. "Naumkeag," *Architectural Digest* 49 (September 1992): 26–34.

Greene, Elaine. "Extraordinary Ordinary: Venturi, Rauch and Scott Brown Design a

Year-Round Hamptons Retreat," *House and Garden* 156 (May 1984): 158–65, 242.

Griffin, Rachael, and Sarah Munro. *Timberline Lodge* (Portland, OR: Friends of Timberline, 1978).

―――. *Timberline Lodge: A Guided Tour*, 2d ed. (Portland, OR: Friends of Timberline, 1991).

Hasbrouck, Wilbert R. "The Earliest Work of Frank Lloyd Wright," *Prairie School Review* 7 (1970): 14–16.

Hitchcock, Henry-Russell. *The Architecture of H. H. Richardson and His Times*, rev. ed. (Cambridge, MA: MIT Press, 1966).

Hurley, Marianne R. "The California Churches of Ernest A. Coxhead: The Legacy of Transplanted Forms," master's thesis, Department of Art History, University of Oregon, 1998.

Kantor, Ann Halpenny. "The Hotel del Coronado and Tent City," in R. G. Wilson, ed., *Victorian Resorts and Hotels* (Philadelphia: Victorian Society in America, 1982).

Kotas, Jeremy. "Ernest Coxhead," in Robert Winter, ed., *Toward a Simpler Way of Life: The Arts and Crafts Architects of California* (Berkeley, CA: University of California Press, 1997).

Ladestro, Debra. "Shingle Style Classic Rebuilt," *Progressive Architecture* (June 1989): 23, 25.

Lawton, Thomas, and Linder Merrill. *Freer: A Legacy of Art* (Washington, D.C.: Smithsonian Institution, 1993).

Longstreth, Richard. *On the Edge of the World: Four Architects in San Francisco at the Turn of the Century* (New York and Cambridge, MA: The Architectural History Foundation and MIT Press, 1983).

Makinson, Randell L. *Greene & Greene: Architecture as Fine Art* (Salt Lake City: Peregrine Smith Books, 1977).

―――. "Greene and Greene," in Esther McCoy, *Five California Architects* (New York: Praeger Publishers, 1975).

―――. *Greene & Greene, David B. Gamble House, Pasadena, California, 1908* (Tokyo: A. D. A. Edita, 1984).

Manson, Grant C. *Frank Lloyd Wright to 1910: The First Golden Age* (New York: Reinhold Publishing Corporation, 1958).

Matthews, Henry. *Kirtland Cutter, Architect in the Land of Promise* (Seattle: University of Washington Press, 1997).

O'Gorman, James F. *H. H. Richardson: Architectural Forms for an American Society* (Chicago: University of Chicago Press, 1987).

―――. *Living Architecture: A Biography of H. H. Richardson* (New York: Simon & Schuster, 1997).

Ochsner, Jeffrey Karl. *H. H. Richardson: Complete Architectural Works* (Cambridge, MA: MIT Press, 1982).

Philadelphia Museum of Art. *Philadelphia: Three Centuries of American Art* (Philadelphia: Philadelphia Museum of Art, 1976).

Reed, Roger. *A Delight to All Who Know It: The Maine Summer Architecture of William R. Emerson* (Portland, ME: Maine Citizens for Historic Preservation, 1995).

Reid, James M. "The Building of the Hotel del Coronado," quoted by Ann Halpenny Kantor, "The Hotel del Coronado and Tent City."

Roper, Laura Wood. *FLO: A Biography of Frederick Law Olmsted* (Baltimore: The Johns Hopkins University Press, 1973).

Roth, Leland M. *A Concise History of American Architecture* (New York: Harper & Row, 1979).

———. *The Architecture of McKim, Mead & White, 1870–1920: A Building List* (New York: Garland Publishing, 1978).

———. *McKim, Mead & White, Architects* (New York: Harper & Row, 1983).

Rueda, Luis F. *Robert A. M. Stern: Building and Projects, 1981–1985* (New York: Rizzoli, 1986).

Sausalito Woman's Club. *Sausalito Woman's Club: Our First Fifty Years* (Sausalito: Sausalito Woman's Club, 1976).

———. *The Sausalito Woman's Club: A Julia Morgan Landmark* (Sausalito: Sausalito Woman's Club, 1988).

Scully, Vincent. *The Architecture of the American Summer: The Flowering of the Shingle Style* (New York: Rizzoli, 1989).

———. *The Shingle Style* (New Haven: Yale University Press, 1955).

———. *The Shingle Style Today: or, The Historian's Revenge* (New York: Braziller, 1974).

———. *The Shingle Style: Architectural Theory and Design from Richardson to the Origins of Wright* (New Haven: Yale University Press, 1955).

Sheldon, George W. *Artistic Country Seats*, vol. 1 (New York, 1886): 22–27. Note that, as is the case with numerous illustrations in Sheldon's book, the gelatine plate of the front of the Bell house was printed backwards.

Silver, Nathan. "Joseph Esherick," in *Contemporary Architects*, rev. ed. (New York: St. James Press, 1994).

Sorell, Susan Karr. "Silsbee: The Evolution of a Personal Architectural Style," *Prairie School Review* 7 (1970): 5–13.

Sherman, Joe. *The House at Shelburne Farms: The Story of One of America's Great Country Estates* (Middlebury, VT: Paul S. Eriksson, 1992).

Steege, Gwen W. "James William Reid," *Macmillan Encyclopedia of Architects* (New York: The Free Press/Collier Macmillan Publishers, 1982).

Stevens, II, John Calvin, and Earle G. Shettleworth, Jr. *John Calvin Stevens: Domestic Architecture, 1890–1930* (Portland, ME: Great Portland Landmarks, Inc., 1995).

Tschirch, John R. "The Isaac Bell House: An Architectural History," *Newport Gazette* 138 (Winter 1996): 2–12.

Van Rensselaer, May King. *Newport: Our Social Capital* (Philadelphia: Lippincott, 1905).

Vogel, Carol. "Architecture: Jaquelin Taylor Robertson," *Architectural Digest* 37 (December 1980): 134–39.

———. "Private House in Eastern Long Island, New York," *Architectural Record* 169 (May 1981, "Record Houses of 1981"): 78–81.

Ward, Ellen MacDonald. "Felsted," *Down East Magazine* (February, 1994): 34–37, 63–65.

White, Samuel. *The Houses of Stanford White* (New York: Rizzoli, 1998).

Williamson, Harold F. *The Growth of the American Economy*, 2d ed. (Englewood Cliffs, NJ: Prentice-Hall, Inc., 1951).

Wilson, Richard Guy. *McKim, Mead & White, Architects* (New York: Rizzoli, 1983).

Woodbridge, Sally, ed. *Bay Area Houses*, new ed. (Salt Lake City: Peregrine Smith Books, 1988).

———. *Bernard Maybeck: Visionary Architect* (New York: Abbeville Press, 1992).

Wright, Frank Lloyd. *An Autobiography* (New York: Longmans, Green and Co., 1932).

Notes

Detail of Sagamore Hill.

1. The publication of the Berkeley house photograph is discussed in L. M. Roth, ed., *America Builds: Source Documents in American Architecture and Planning* (New York, 1983): 232–34. The text accompanying the photograph is included.

2. Ibid.

3. For McKim's earliest independent work see Richard Guy Wilson, "The Early Work of Charles F. McKim: Country House Commissions," in *Winterthur Portfolio* 14 (Autumn 1979): 235–67.

4. See Andrew Saint, *Richard Norman Shaw* (New Haven and London: Yale University Press, 1976): 24–52, and C. Murray Smart, Jr., "Richard Norman Shaw," in Randall J. Van Vynckt, ed., *International Dictionary of Architects and Architecture* 1 (Detroit, Michigan: St. James Press, 1993): 814–15.

5. McKim's photographic portfolio, titled "Old Newport Houses, 1875," is mentioned in Walter Knight Sturges, "Arthur Little and the Colonial Revival," *Journal of the Society of Architectural Historians* (May 1973): 147–63. Sturges notes that McKim gave the portfolio to William Dean Howells, an early client, perhaps to illustrate some details for Howells's new house; Howells later presented the portfolio to the Society for the Preservation of New England Antiquities (SPNEA), where it can be found today.

6. The reminiscences of Mead, recalling the firm's early days, seem to have disappeared but are quoted at length in Charles Moore, *The Life and Times of Charles Follen McKim* (Boston: Houghton Mifflin Company, 1929): 41.

7. The Gray house is discussed and illustrated with a plan and wood-block-engraved interior vignettes in Mariana Griswold Van Rensselaer, "American Country Dwellings," Part 2, *The Century Magazine* 32 (June 1886): 206–20; this is reprinted (minus some of the illustrations) in David Gebhard, ed., *Accents As Well As Broad Effects: Writings on Architecture, Landscape, and the Environment, 1876–1925* [by Mariana Griswold Van Rensselaer] (Berkeley: University of California Press, 1996): 243–59. An edited version of Mrs. Van Rensselaer's essay, with the Gray House plan and an interior of the Drawing Room, is reprinted in L. M. Roth, *America Builds*, 246–50.

8. See William B. Rhoads, *The Colonial Revival* (New York: Garland Publishing, 1977): 1:50, 346; 2:585, n. 5, 790, n. 10. Another reference to Poore's antiquarian collecting is in William B. Hosmer, Jr., *Presence of the Past* (New York: G. P. Putnam's Sons, 1965): 211–12. Poore had sections of panelling from the Old Province House and from the John Hancock Mansion, a large collection of early American furniture, and apparently sections of stucco work with embedded glass. Poore was well known as a genial host who loved to receive visitors.

9. For Aestheticism see The Metropolitan Museum of Art, *In Pursuit of Beauty: Americans and the Aesthetic Movement* (New York: Rizzoli, 1986), produced by the

museum in conjunction with an exhibition, and Mark Girouard, *Sweetness and Light: The "Queen Anne" Movement, 1860–1900* (Oxford: Clarendon Press, 1977).

10. For Edward S. Morse, his book, and his influence in the United States, see Kevin Nute, *Frank Lloyd Wright and Japan* (New York: Van Nostrand, 1993). Morse's book is kept in print by Dover Publications, Mineola, New York.

11. Maybeck's scrapbook photograph is reproduced in Richard Longstreth, *On the Edge of the World: Four Architects in San Francisco at the Turn of the Century* (New York and Cambridge, MA: The Architectural History Foundation and MIT Press, 1983): 119. Longstreth makes no reference to the sheathing material but mentions only the picturesque qualities that may have influenced Polk. It is evident that Polk had access to the scrapbooks, as Maybeck was also practicing in San Francisco.

12. Nearly all the Shingle Style buildings mentioned here are illustrated by the architects' original perspective drawings collected in Vincent Scully, *The Architecture of the American Summer* (New York: Rizzoli, 1989).

13. The broad triangular gable houses by McKim, Mead & White, culminating in the William G. Low house of 1886–87, are discussed as a group in Richard Guy Wilson, "The Big Gable," *Architectural Review* 176 (August 1984): 52–57.

14. For the Newsom brothers and family see David Gebhard, "Samuel Newsom and Joseph Cather Newsom," *Macmillian Encyclopedia of Architects* (New York: The Free Press/Collier Macmillan Publishers, 1982), 3:292–98, and David Gebhard et al., *Samuel and Joseph Cather Newsom: Victorian Architectural Imagery in California, 1878–1908* (Santa Barbara, CA: University of California-Santa Barbara Art Museum, 1979).

15. See M. Shellenbarger, K. Lakin, and L. M. Roth, *The Architecture and Teaching of Ellis F. Lawrence* (Eugene, OR: Museum of Art and School of Architecture and Allied Arts, 1989) and Meredith Clausen, *Pietro Belluschi: Modern American Architect* (Cambridge, MA: MIT Press, 1994).

16. For the Silsbee shingled houses in Edgewater, see one example reproduced in L. M. Roth, *A Concise History of American Architecture* (New York: Harper & Row, 1979): 153, fig. 132. See also the two essays on Silsbee by Susan K. Sorell in *Prairie School Review* 7 (Fourth Quarter, 1970): 5–13, 17–21, which include several reproductions.

17. For Ernest Coxhead's career, see Longstreth, *On the Edge of the World*.

18. For the complexities of Willis Polk's career see also Longstreth, *On the Edge of the World*.

19. Ibid.

20. These other early shingled buildings by Brown are reproduced in Scully's *The Architecture of the American Summer*.

21. Ibid. See also K. Cardwell, *Bernard Maybeck: Artisan, Architect, Artist* (Salt Lake City: Peregrine Smith Books, 1983) and Sally B. Woodbridge, *Bernard Maybeck, Visionary Architect* (New York: Abbeville Press, 1992). Cardwell reproduces the E. H. Johnson house drawing from *American Architect and Building News*, August 1888.

22. Queen Anne was the style developed by Shaw immediately following what was called "Old English"; using classical details, Queen Anne was based more on seventeenth-century sources, whereas "Old English" by comparison derived largely from sixteenth-century sources.

23. See the essay on Scully by James Stevenson, "Profiles [Vincent Scully]: What Seas, What Shores," in *The New Yorker* 55 (February 18, 1980): 43–48, 52–59, 60–61.

Detail of the stair landing at Shelburne House.

INDEX

ACKNOWLEDGMENTS

"The Airplane House"

The photography of this book was made possible by the generosity of spirit of the owners, curators, and caretakers of the structures portrayed. The owners of private houses offered their cheerful hospitality and accommodated the rigors of architectural photography. Extraordinary access and assistance were also offered by the following individuals at these locations: Linda Johnson, public relations manager, International Tennis Hall of Fame (the Newport Casino); Monique Panaggio, public relations director, the Preservation Society of Newport County (Kingscote and the Isaac Bell House); Susan Sarna of the National Park Service at Sagamore Hill; Marsha Goodwin and Ann Clifford for the City of Waltham at Stonehurst; Mark Baer, historic house administrator for the Trustees of Reservations at Naumkeag; Hilary Sunderland at Shelburne Farms; Lauren Ash Donoho, public relations director, the Hotel del Coronado; the Reverend David Miller at St. John's Church, Petaluma; Edward R. Bosley III, director, and the staff of the Gamble House; Gardner Miller, dedicated caretaker of "The Airplane House"; Chipper Roth, the gracious tenant of the John Galen Howard House; Millie Amis at the Sausalito Woman's Club; and Sara Gokke at Timberline Lodge.

The resources of the Avery Architectural and Fine Arts Library at Columbia University were invaluable in researching the Shingle Style and related developments, and in identifying and locating extant structures. Librarians Janet Parks and Anne Roure provided period illustrations for the Introduction. Other period photographs were provided by M. Joan Youngken of the Newport Historical Society, Joanne S. Dunlap of the Whitehall Museum House, and Richard Longstreth of The George Washington University. Earle G. Shettleworth, Jr. of the Maine Historic Preservation Commission offered his knowledge of structures in Maine. Pamela Cromey of the Tuxedo Historical Society provided information on the work of Bruce Price at Tuxedo Park. Nancy M. U. Meade shared memories of her grandfather, Charles Alonzo Rich.

John Tucker of Norfleet Press contributed discerning and unflappable direction of this project from its inception. Charlotte Staub brought both organization and canny intuition to the design of the book. At Harry N. Abrams, Inc., editor Diana Murphy saw the book into print with great faith in the vision of its creators. Julia Gaviria's eye for detail and accuracy yielded innumerable refinements to the book.

BRET MORGAN
LELAND M. ROTH